Lecture Notes in Computer Science 13034

More information about this subseries at http://www.springer.com/series/7407

Jin Akiyama · Reginaldo M. Marcelo ·
Mari-Jo P. Ruiz · Yushi Uno (Eds.)

Discrete and Computational Geometry, Graphs, and Games

21st Japanese Conference, JCDCGGG 2018
Quezon City, Philippines, September 1–3, 2018
Revised Selected Papers

Editors
Jin Akiyama
Tokyo University of Science
Tokyo, Japan

Reginaldo M. Marcelo
Ateneo de Manila University
Quezon City, Philippines

Mari-Jo P. Ruiz
Ateneo de Manila University
Quezon City, Philippines

Yushi Uno
Osaka Prefecture University
Osaka, Japan

ISSN 0302-9743 ISSN 1611-3349 (electronic)
Lecture Notes in Computer Science
ISBN 978-3-030-90047-2 ISBN 978-3-030-90048-9 (eBook)
https://doi.org/10.1007/978-3-030-90048-9

LNCS Sublibrary: SL1 – Theoretical Computer Science and General Issues

This Springer imprint is published by the registered company Springer Nature Switzerland AG
The registered company address is: Gewerbestrasse 11, 6330 Cham, Switzerland

Preface

This volume consists of selected papers presented at the 21st Japan Conference on Discrete and Computational Geometry, Graphs, and Games (JCDCGGG 2018) held on September 1–3, 2018, at Ateneo de Manila University in Quezon City, Philippines. The conference is part of the series which started out as the Japan Conference on Discrete and Computational Geometry (JCDCG) in 1997, organized by Jin Akiyama (of Tokai University at the time) and his colleagues. Conferences in the series have been held annually, except for the years 2008 and 2020, usually in Japan, although some were hosted by other Asian countries such as the Philippines, Thailand, China, and Indonesia. Graphs became one of the conference topics soon after the series started, while Games were included recently.

We are very grateful to the more than 70 referees who reviewed a total of 30 submitted papers. We thank the authors for their patience and cooperation during the back-and-forth process of comments and revisions initiated by the editors and the referees, especially since the process slowed down at the onset of the COVID-19 pandemic late in 2019 through 2020. Finally, we thank the Springer team for the ready assistance, as we learned and complied with the publication requirements.

August 2021

Jin Akiyama
Reginaldo M. Marcelo
Mari-Jo P. Ruiz
Yushi Uno

Organization

Conference Chairs

Jin Akiyama	Tokyo University of Science, Japan
Mari-Jo P. Ruiz	Ateneo de Manila University, Philippines

Organizing Committee

Agnes Garciano	Ateneo de Manila University, Philippines
Hiro Ito	University of Electro-Communications, Japan
Mikio Kano	Ibaraki University, Japan
Reginaldo M. Marcelo	Ateneo de Manila University, Philippines
Toshinori Sakai	Tokai University, Japan
Jumela Sarmiento	Ateneo de Manila University, Philippines
Yushi Uno	Osaka Prefecture University, Japan

Program Commitee

Reginaldo M. Marcelo	Ateneo de Manila University, Philippines
Ian June Garces	Ateneo de Manila University, Philippines
Agnes Garciano	Ateneo de Manila University, Philippines
Mikio Kano	Ibaraki University, Japan
Toshinori Sakai	Tokai University, Japan
Mark Anthony Tolentino	Ateneo de Manila University, Philippines
Yushi Uno	Osaka Prefecture University, Japan

Secretariat

Mark Anthony Tolentino
Jonn Angel Aranas
Engel John Dela Vega
Jakov Dumbrique
Janree Ruark Gatpatan
Luis Silvestre, Jr.
Immanuel Gabriel Sin

Logistics

Jude Buot
Mark Loyola
Eurlyne Domingo
Joanna Cardiño
Patrick John Fernandez
Edwin Morales
Edrian Neverio
Lu Christian Ong
Renzo Roel Tan
Lean Franzl Yao
Anthony Zosa

Finance

Jumela Sarmiento
Editha Bagtas

Contents

On Geometric Graphs on Point Sets in the Plane

Jorge Urrutia(✉)

Instituto de Matemáticas, Universidad Nacional Autónoma de México, Mexico City, Mexico
urrutia@matem.unam.mx

Abstract. A graph whose vertex set is a set P of points in the plane and whose edges are line segments joining pairs of elements of P is called a *geometric graph*. In this paper we survey several results on geometric graphs on colored point sets. Of particular interest are bicolored point sets $P = R \cup B$ in which the elements of P can considered to be colored red or blue. We will pay particular attention to perfect matchings, spanning trees and paths whose vertex sets are colored point sets. In the last section of this paper we give some results on point sets whose elements are labelled with the integers $\{1, \ldots, n\}$ such that different elements of P receive different labels.

1 Introduction

In this paper we will survey several results on geometric graphs on colored point sets; that is, point sets in general position $P = C_1 \cup \ldots \cup C_k$, where we may assume that the elements of C_i are colored with color i, $1 \leq i \leq k$. When we use two colors, we usually call $P = R \cup B$ a bicolored point set. The elements of R and B are usually called the red and the blue points of P. We will say that a bicolored point set is a *balanced bicolored* set if $|R| = |B|$.

Many of the results included in this survey originated in editions of the Japanese Conferences on Discrete and Computational Geometry, and in many cases were the result of our collaboration with Japanese friends and colleagues.

2 Perfect Matchings on Bicolored Point Sets

Let $P = R \cup B$ be a balanced bicolored point set. A perfect *red–blue* matching M of P is a set of n disjoint pairs of points $(b_1, r_1), \ldots, (b_n, r_n)$ such that $b_i \in B$ and $r_i \in R$, $i = 1, \ldots, n$. We say that M is *plane* if the n line segments joining r_i to b_i are mutually disjoint; these line segments will be called the *edges* of M. See Fig. 1. The first result, which I learned many years ago from Prof. Jin Akiyama and is due to Atallah [12], is the following.

Research supported in part by PAPIIT grant IN102117 Universidad Nacional Autónoma de México.

J. Akiyama et al. (Eds.): JCDCGGG 2018, LNCS 13034, pp. 1–17, 2021.
https://doi.org/10.1007/978-3-030-90048-9_1

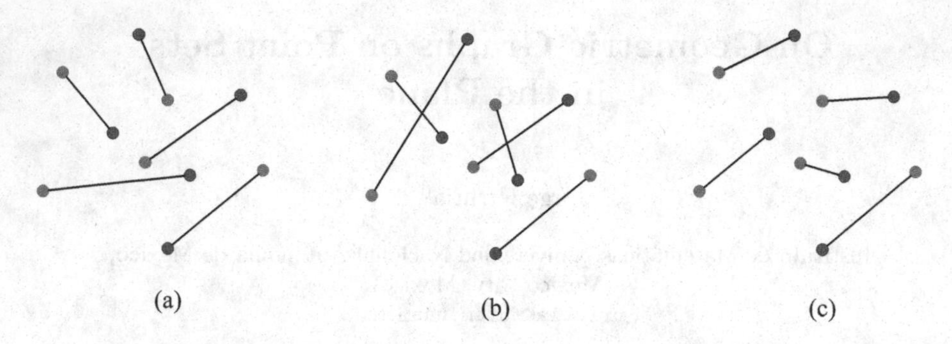

Fig. 1. (a) A plane bichromatic matching (b) A non-plane bichromatic matching. (c) Uncrossing two pairs of edges. (Color figure online)

Theorem 1. *Every balanced bicolored point set always has a plane red–blue perfect matching.*

It is easy to see that the minimum weight perfect red–blue matching M, where the weight of M is the sum of the lengths of its edges, is plane. Indeed, if a pair of edges of M cross, we can replace them by two non-crossing red–blue edges such that the sum of their lengths is smaller than the original pair, contradicting the claim that M was of minimum weight. See Fig. 1(c). Computing a minimum weight plane matching for balanced bicolored point sets can be done in $O(n \log n)$ time using the linear time algorithm of Lo et al. [26] to compute a ham-sandwich cut.

An extension of Theorem 1 to higher dimensions by Akiyama and Alon was obtained in [6]. Let A be a set of dn points in general position in R^d and let $A = A_1 \cup \ldots \cup A_d$ be a partition of A into d pairwise disjoint sets, each consisting of n points. Then there are n pairwise disjoint $(d-1)$-dimensional simplices, each containing precisely one vertex from each A_i, $1 \leq i \leq d$.

3 Alternating Paths

Another early result on balanced bicolored point sets is that of finding *simple alternating paths.* Let $P = R \cup B$ be a balanced bicolored point set. A *simple alternating path* of P is a *simple polygonal path* $\mathcal{W}$, all of whose vertices are in P, with each edge in $\mathcal{W}$ having a red and a blue vertex. See Fig. 2. Most of the results on simple alternating paths that we know have to do with balanced bicolored point sets P in *convex position*; that is, the elements of P are the vertices of a convex polygon; see Fig. 3.

A path whose vertices are all of the vertices of a point set P is called a *spanning* path. There are bicolored point sets that admit no simple alternating spanning path. The following point set with $2n$ points obtained by Erdős [24] admits no simple alternating paths with more than $\frac{3n}{2} + 1$ elements: place $\frac{n}{2}$ consecutive red points on a circle, followed by $\frac{3n}{4}$ consecutive blue points, followed by $\frac{n}{2}$ red points, followed by $\frac{n}{4}$ blue points, $n = 4m$, $m \geq 4$. To the

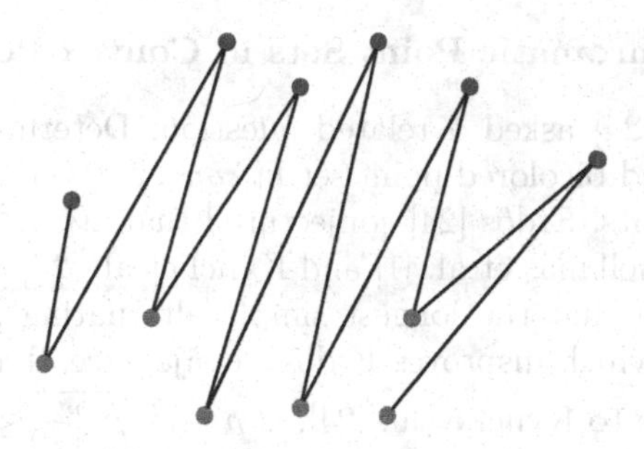

Fig. 2. A simple alternating path of a bicolored point set. (Color figure online)

best of my knowledge, simple alternating paths of balanced bicolored point sets were first studied by Akiyama and Urrutia in [8]. In that paper, they obtained a dynamic programming quadratic-time algorithm to determine if a balanced bicolored point set in convex position has a simple alternating spanning path; see Fig. 3. It is easy to see that the algorithm given in [8] for balanced bicolored point sets in convex position can be modified to produce the longest simple alternating path that a point set P may have, or how many spanning alternating paths P has; see Abellanas *et al.* [1]. The following problem is, as far as we know, open.

Problem 1. Is there a polynomial time algorithm to decide if a balanced bicolored point set in general position has a simple spanning alternating path, or is this problem NP-hard?

Our suspicion is that deciding if a balanced bicolored point set has a simple alternating path is NP-hard. Abellanas et al. [2] proved that any balanced bicolored point set P, such that the red and blue points of P are separated by a line, always has a simple alternating spanning path. Using the ham-sandwich theorem together with the result of Abellanas et al. [2], it follows that any balanced bichromatic point set with $2n$ elements always has a simple alternating path that covers at least half of the elements of P [2]. A challenging conjecture is the following.

Conjecture 1. There is a constant $c > 1$ such that any balanced bicolored point set P with $2n$ elements has a simple alternating path covering at least cn elements of P.

Conjecture 1 is open even for balanced bicolored point sets in convex position. If we allow edges to have few intersections, then the following results, proved Kaneko el al. [22] and Claverol et al. [15], are true.

Theorem 2 [22]. *Any balanced bicolored point set with* $2n$ *elements has a spanning alternating Hamilton cycle containing at most* $n - 1$ *edge crossings*

Theorem 3 [15]. *Any balanced bicolored point set in general position has a spanning alternating Hamilton cycle* C *in which every edge is crossed by at most one other edge of* C.

3.1 Balanced Bichromatic Point Sets in Convex Position

Around 1989 Erdős [24] asked a related question: Determine the largest $\ell(n)$ such that any balanced bicolored point set *in convex position* has an alternating path with $\ell(n)$ elements. Erdős [24] conjectured that $\ell(n) < \frac{3n}{2} + 2$.

Independently, Abellanas et al. [1] and Kynčl et al. [24] found bicolored balanced point sets such that the longest simple alternating path in the set has $\frac{4n}{3} + c\sqrt{n}$ elements, which disproves Erdős's conjecture. The best lower bound known to us now, due to Kynčl et al. [24], is $n + c'\sqrt{\frac{n}{\log n}} \leq \ell(n)$.

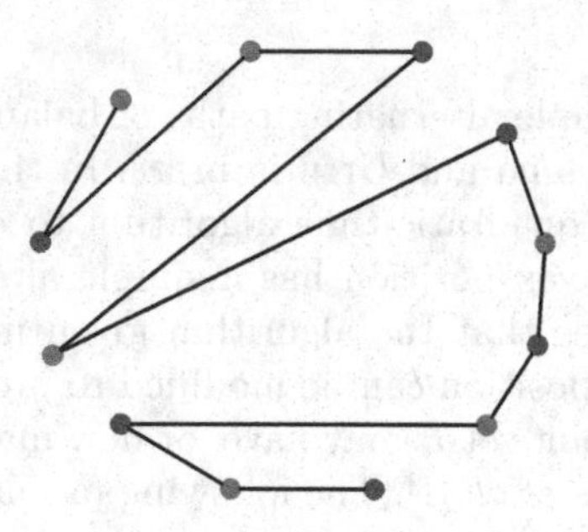

Fig. 3. An alternating path of a balanced bicolored point set in convex position. (Color figure online)

We now present another proof that any balanced point set in convex position has a simple alternating path that covers at least half of its elements. Using the idea behind this proof, Merino et al. [28] proved that any balanced tricolored point set in convex position, and with $3n$ elements, always has a *simple alternating path* with $2n + 1$ elements; that is, a simple path such that the vertices of each edge have different colors. Their bound is tight, as there are point sets with $3n$ elements that have no alternating paths with more than $2n + 1$ vertices.

Consider a balanced bicolored point set in convex position, and let p be any element of P; see Fig. 4. Suppose without loss of generality that p is the leftmost point in P, and that it is a blue point. In a greedy way we will choose two simple alternating paths starting at p as follows: Let $b_1 = p$. Starting from b_1, let r_1 be the first red point we meet when we traverse the convex hull of P in the clockwise direction. Choose b_2 as follows: starting from b_1 traverse the convex hull of P counterclockwise until we find the first blue point, call it b_2; see Fig. 4(a). We can obtain $r_2, b_3, \ldots$ recursively until we have to stop at a point q. Repeat the previous process, starting at b_1, but this time instead of traversing the convex hull of P in the clockwise direction, traverse it counter-clockwise until we find a red point, call it r'_1. The second point in our path is now chosen by traversing the boundary of P in the clockwise direction until we find a blue point, call it b'_2. We continue again recursively until we have to stop; see Fig. 4(b). It is easy to see that we obtain two simple alternating paths that end at the same point

q, such that their union covers all of the elements of P; see Fig. 4(c). It follows now one of the paths thus obtained contains at least half of the elements of P.

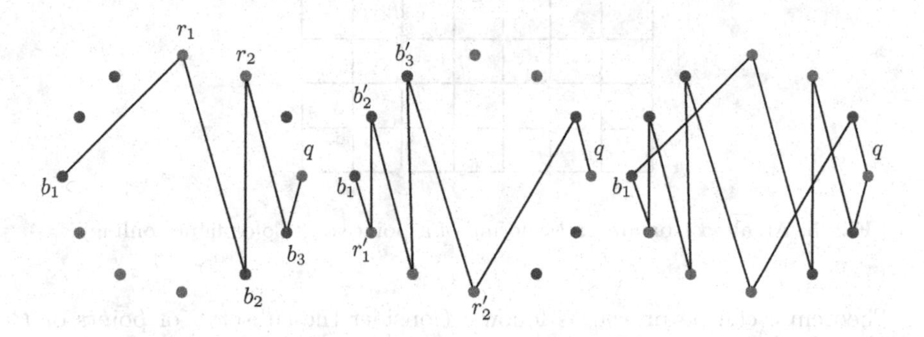

Fig. 4. Choosing two alternating paths. (Color figure online)

We conclude this section with the following open problem.

Problem 2. Let P be a point set in general position with $3n$ elements, n colored red, n colored *blue*, and n colored green. Is it true that P always has a simple alternating path covering at least $2n \pm c$ elements of P?

In this section we have dealt mostly with problems in which we seek *simple* alternating paths. In Sects. 5, 6 and 7 we study problems on geometric graphs whose edges are allowed to intersect. This opens up a set of nice problems.

4 Balanced Colorings of Lattice Point Sets

Let $\mathcal{L}$ be the set of points in the plane with integer coordinates, and $P \subset \mathcal{L}$. P is called a *lattice point set.* An *almost balanced k-coloring* of P is a k-coloring of the elements of P such that in any line ℓ parallel to either of the coordinate axes, the difference (in absolute value) between the number of points of P colored i and colored j is at most one, $1 \leq i < j \leq k$, see Fig. 5. Akiyama and Urrutia [7] proved the following result:

Theorem 4 [7]. *Let $P \in \mathcal{L}$ with n elements, and k an integer, $1 \leq k \leq n$. Then there always exists an almost balanced k-coloring of P.*

As it turns out, Theorem 4 is a consequence of a following well known result on edge colorings of bipartite graphs.

Theorem 5. *Let G be a bipartite graph with maximum degree k. Then the edges of G can be colored with k colors in such a way that any two edges incident to a vertex of G have different colors.*

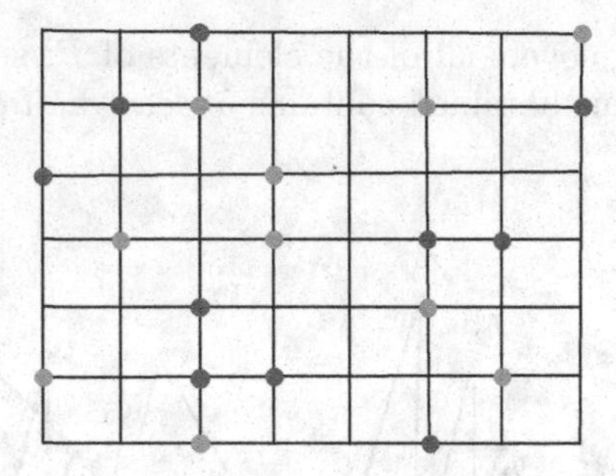

Fig. 5. An almost balanced 2-coloring of a point set. (Color figure online)

Theorem 4 can be proved as follows: Consider the subset S of points of P on a horizontal (respectively, vertical) line ℓ. Suppose that S has $m = rk + s$ for some k and s. Split the elements of S into r disjoint subsets with k elements, and if $s > 0$, one subset with s elements; see Fig. 6. These sets will be called *horizontal* (respectively, *vertical*) subsets.

Next construct a bipartite graph G whose vertices are the horizontal and vertical subsets of P, two of which are adjacent subsets if they intersect. G is a bipartite graph with maximum degree k, and by Theorem 5 its edges can be colored with k colors such that edges incident to a vertex receive different colors. Observe that there is a one-to-one mapping between the edges of G and the points in P, and thus the edge coloring of G induces a coloring of the points of P. It is easy to see that this coloring is almost balanced, as the elements in each horizontal or vertical subset of P receive different colors; see Fig. 6.

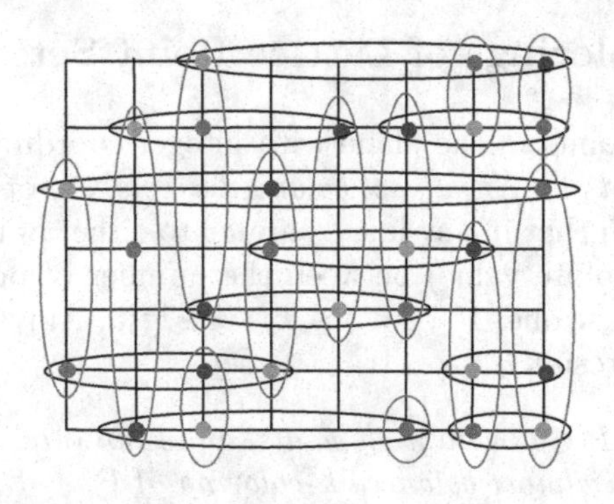

Fig. 6. Obtaining an almost 3-balanced coloring of a lattice point set. (Color figure online)

This result was generalized to higher dimensions by Biedl et al. [14]. Among other results, they proved that lattice point sets in dimension d can always be k-colored in such a way that for any two colors a and b, the difference between the number of points colored a and the number of points colored b is at most

$4d - 3$, so for 2-colorings the difference is at most $max\{2d - 3, 2\}$. Furthermore they proved that for higher dimensions, the problem of deciding if a lattice point set has a balanced 2-coloring is NP-complete.

5 Monochromatic Spanning Trees with Few Intersections

Given a point set Q, a Q-tree is a spanning geometric tree whose vertices are the points in Q, and no two of its edges intersect except perhaps at their endpoints. Tokunaga studied the following problem: Let $P = R \cup B$ be a not necessarily balanced bicolored point set. Find a B-tree and an R-tree that intersect as few times as possible; see Fig. 7.

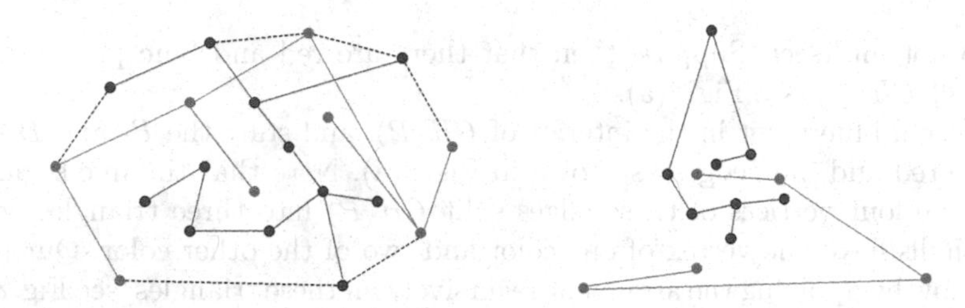

Fig. 7. Two R- and B-trees that intersect the fewest number of times. (Color figure online)

Let $g(R, B)$ denote the number of bichromatic edges on the convex hull $CH(P)$ of P. Tokunaga [33] proved the following.

Theorem 6 [33]. *Let $f_T(R, B)$ denote the minimum number of intersection points of an R-tree and a B-tree. Then*

$$f_T(R, B) = \max\left\{\frac{g(R, B)}{2} - 1, 0\right\}.$$

In other words, $f_T(R, B)$ depends only on the number of bichromatic edges on the boundary of the convex hull of P, regardless of the position and number of points of P in the interior of the convex hull of P.

We prove here the case when the convex hull of P contains exactly three vertices, two of one color and one of the other. Note that in this case $g(R, B) = 2$, and $f_T(R, B) = 0$; that is, there must be an R- and a B-tree whose edges do not intersect. Theorem 6 follows easily from this case by splitting the interior of the convex hull of P into a set of triangles with disjoint interiors such that their vertices belong to the convex hull of P.

Consider a bicolored point set whose convex hull has three points, two red and one blue; see Fig. 8. Note that if all the points in the interior of $CH(P)$ are colored with the same color, then we can obtain an R- and a B-tree such that

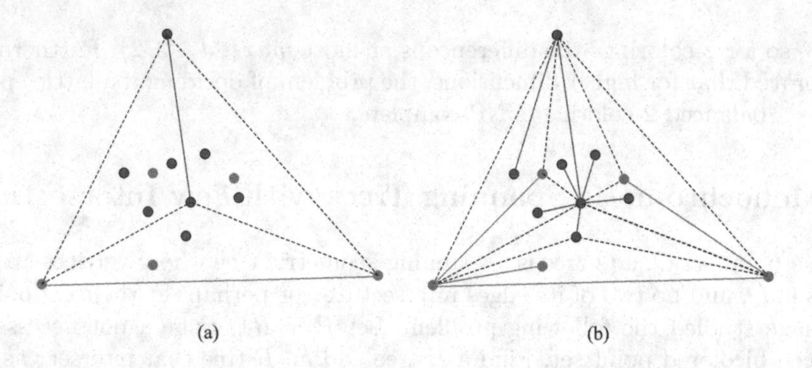

Fig. 8. Illustrating the recursive proof of Theorem 6. (Color figure online)

they do not intersect. Suppose then that there are red and blue points in the interior of $CH(P)$; see Fig. 8(a).

Choose a blue point in the interior of $CH(P)$, and start the R- and B-trees with the red and blue edges as shown in Fig. 8(a). Note that the line segments joining the four vertices of these edges split $CH(P)$ into three triangles whose convex hulls have one vertex of one color and two of the other color. Our proof follows now by applying the argument recursively on these triangles; see Fig. 8(b).

An R-cycle (resp. B-cycle) is a simple hamiltonian cycle that covers all the elements of R (resp. B). The following conjecture by Tokunaga is still open.

Conjecture 2 [33]. There exists a pair of an R-cycle and a B-cycle such that each edge of either one of the two cycles intersects the other cycle at most twice.

Tokunaga's result does not generalize to point sets colored with three or more colors. However, some tight bounds on the intersection number of matchings, spanning trees and paths have been obtained in Kano et al. [23], Leaños et al. [25], Merino et al. [27], and Suzuki [32]. Some of these results will be reviewed more carefully in Sect. 7.

6 Mutually Crossing Paths

Most results on geometric graphs involve the study of plane structures; e.g., planar bichromatic matchings, alternating paths, etc. In this section we will recall some results in which we want to do the opposite; that is, we will be interested in obtaining geometric graphs with many intersections. One of the first papers we know in this direction was Aronov et al. [11]. In that paper the authors proved that any set P of n points, in general position in the plane, has a matching with $\sqrt{\frac{n}{12}}$ elements, all of which cross each other; they call such matchings *crossing* matchings. Pach and Solymoshi [30] then proved that if a point set with $n = 2m$ points has exactly m halving lines, then it has a crossing perfect matching. Recently Pach et al. [29] proved that any balanced bicolored point set always has a crossing matching with at lest $n^{1-o(n)}$ elements.

A *crossing family of k-paths* on P is a set of vertex-disjoint k-paths such that any two of them cross. The next result was proved by Álvarez Rebollar et al. [9].

Theorem 7 [9]. *For any point set P with n points in the plane in general position there always exists a crossing family of vertex-disjoint 3-paths of size $\lfloor n/4 \rfloor$.*

Proof. Assume that P has $4m$ elements. Draw two vertical auxiliary lines, $\mathcal{L}_1$ and $\mathcal{L}_2$, such that the plane is divided into three sections, $\mathcal{A}$, $\mathcal{B}$ and $\mathcal{C}$ from left to right, such that $\mathcal{A}$ and $\mathcal{C}$ contain each m points in P, and $\mathcal{B}$ contains $2m$ points of P; see Fig. 9. Next consider an arbitrary perfect matching of P in which every element of P in $\mathcal{A} \cup \mathcal{C}$ is joined to a different element of $\mathcal{B}$. This generates m vertex-disjoint edges that cross $\mathcal{L}_1$ and m different edges that cross $\mathcal{L}_2$, as shown in Fig. 9.

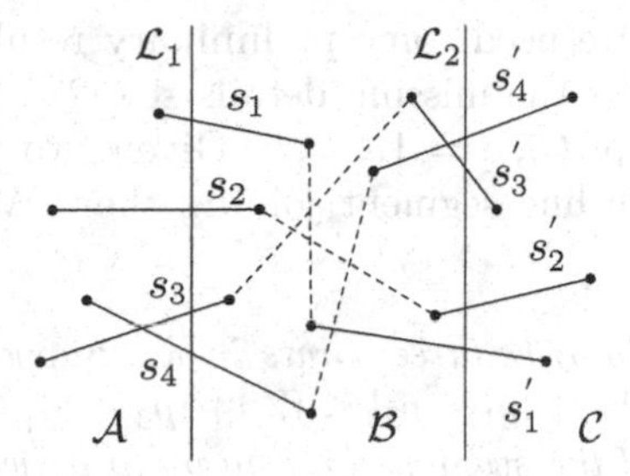

Fig. 9. Constructing a family of crossing 3-paths.

Label the edges that cross $\mathcal{L}_1$ by $s_1, \ldots, s_m$ according to the order in which they intersect $\mathcal{L}_1$ from top to bottom. Label the edges that cross $\mathcal{L}_2$ by $s_1', \ldots, s_m'$ according to the order in which they cross $\mathcal{L}_2$, but from bottom to top. Now join the right endpoint of s_i with the left endpoint of s_i', $i = 1, \ldots, \frac{n}{4}$. It is then easy to see that we obtain a set of crossing 3-paths.

Since any 3-path has four elements of P as vertices, any family of vertex-disjoint 3-paths of P has at most $\lfloor \frac{n}{4} \rfloor$ elements, which shows that our bound is optimal.

In [9] it is also proved that for every point set P of n points in the plane in general position, there exists a Hamiltonian cycle such that the number of crossings between the edges of the cycle is at least $\frac{n^2}{12} - O(n)$. They also prove that there is a set of $\lfloor \frac{n}{6} \rfloor$ vertex-disjoint crossing triangles whose vertices are in P. We close this section with the following conjecture.

Conjecture 3 [9]. Every set of n points in general position in the plane has a crossing family of 2-paths of linear size.

Note that the recent result of Pach et al. [29] on mutually crossing matchings of size $n^{1-o(n)}$ implies the existence of $n^{1-o(n)}$ mutually crossing 2-paths.

7 Colored Matchings with Few Intersections

In this section C_i will denote a point set with $2n$ points, and M_i a plane perfect matching of C_i. We now study the following problem: Let P be a point set such that $P = C_1 \cup \ldots \cup C_k$. Can we find a set of k plane perfect matchings $M_1, \ldots, M_k$, one for each C_i, such that the edges of these matchings do not intersect many times? The following result was proved by Merino et al. [27].

Theorem 8 [27]. *Let P be as above. Then there are k perfect matchings $M_1, \ldots,$ M_k, one for each C_i, such that the number of crossings between the edges of the k matchings is at most*

$$\binom{k}{2}(2n-1).$$

The bound is tight.

To prove Theorem 8 we need some preliminary results, some of which will be given without proof; for the missing details see [27]. Let M_i be a minimum weight perfect matching of C_i, $i = 1, \ldots, k$. Given two points p and q, $|p - q|$ denotes the length of the line segment joining them. We prove the following result [27].

Lemma 1. *Let p_1, p_2 and p_3 be three points in $\Re^2$. Suppose that for some values W and W' we have that $W + |p_1 - p_2| \leq W' + |p_2 - p_3|$. Then if we choose any point p_2' in the interior of the segment joining p_1 to p_2 we have $W + |p_1 - p_2'| <$ $W' + |p_2' - p_3|$. See Fig. 10.*

Let $P = \{p_1, \ldots, p_{2n}\}$. These results then follow from Lemma 1.

Lemma 2. *Let M be a minimum weight perfect matching of a point set Q. Suppose that the edges of M are $\{p_1, p_2\}, \{p_3, p_4\}, ..., \{p_{2n-1}, p_{2n}\}$. Then if for $i = 1, 3, ..., 2n-1$ we choose two points p_i' and p_{i+1}' in $p_i - p_{i+1}$, p_i' closer to p_i than p_{i+1}', then*

$$|p_1' - p_2'| + |p_3' - p_4'| + ... + |p_{2n-1}' - p_{2n}'| < |p_2' - p_3'| + |p_4' - p_5'| + ... + |p_{2n}' - p_1'|.$$

The next result, called the *Shrinking Lemma*, follows; see Fig. 11:

Lemma 3. *Let M be a minimum weight perfect matching of a point set Q. For every edge $\{p_i, p_j\}$ of M, let p_i' and p_j' be two points on the* closed *line segment joining p_i to p_j. Then the set of edges $\{p_i', p_j'\}$ such that $\{p_i, p_j\} \in M$ forms a minimum weight perfect matching of $Q' = \{p_1', ..., p_{2n}'\}$.*

Let M_i and M_j be minimum weight perfect matchings of C_i and C_j.

Theorem 9 [27]. *The intersection graph of the edges of $M_i \cup M_j$ is a forest.*

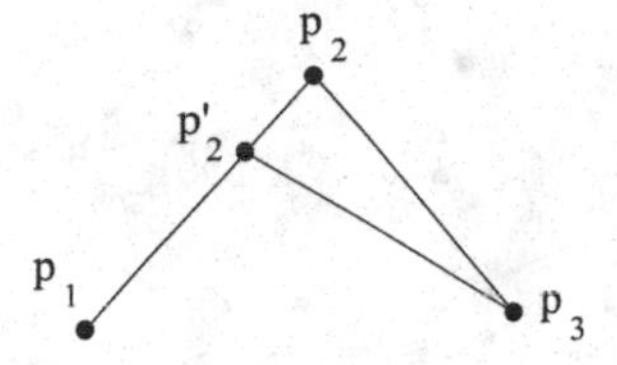

Fig. 10. Shrinking segment $p_1 - p_2$.

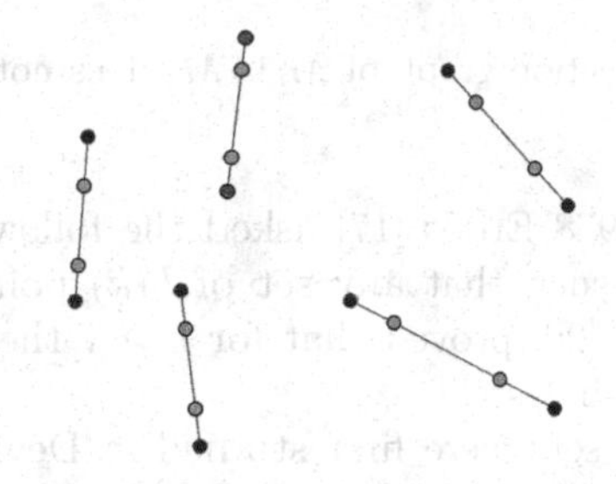

Fig. 11. If the line segments joining the blue pairs of points is a minimum weight perfect matching of the set of blue points, then the short segments joining the green pairs of points also is a minimum weight matching of the set of green points. (Color figure online)

Proof. Suppose that the intersection graph of the edges of $M_i \cup M_j$ is not a forest, and that it contains a cycle C. Clearly this cycle is of even length, and the edges of M_i in C do not cross; see Fig. 12. But by the Shrinking Lemma, the blue (respectively, red) line segments joining the green points that are contained in the blue (resp. red) segments of M_i (respectively, M_j) form a minimum weight matching of the green points. Thus their weights are the same; but by Lemma 2 this is impossible. ∎

If $M_i| = r$ and $|M_j| = s$, from Theorem 9 we have the following lemma.

Lemma 4. *The edges of M_i and M_j intersect at most $r + s - 1$ times.*

Theorem 8 follows by applying Lemma 4 to all pairs of $M_i, M_j, 1 \leq i < j \leq k$.

Kano et al. [23] proved that that given two sets of points R and B, their minimum weight spanning trees (minimum weight hamiltonian cycles) intersect at most a linear number of times; however, determining the maximum number of times such trees intersect is still open.

8 Holes in Point Sets

A *k-hole* of a point set P is a subset of points of P that are the vertices of a convex k-gon Q, such that Q contains no elements of P in its interior. The study of k-holes in point sets was first discussed in a seminal paper by Erdős

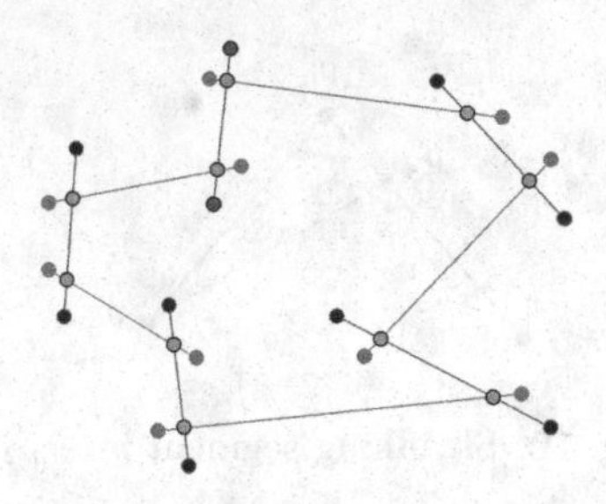

Fig. 12. The intersection graph of $M_i \cup M_j$ does not contain cycles.

and Szekeres in [18]. In 1978 Erdős [17] asked the following question: What is the smallest integer $h(k)$ such that any set of $h(k)$ points in general position contains a k-hole? Horton [20] proved that for $k \geq 7$ there are arbitrarily large point sets containing no 7-holes.

Holes in colored point sets were first studied in Devillers et al. [16], where they proved that any large enough bichromatic point set always has a 3-hole, all of whose vertices have the same color. From here on, these holes will be called *monochromatic empty triangles.* In fact they proved that any large enough bicolored point set always has a linear number of monochromatic empty triangles. They also showed that the so-called Horton point sets can be colored with three colors in such a way that no empty monochromatic triangle exists. A question arising from the results in [16] is the following.

Problem 3. Is it true that any bicolored point set contains a super-linear number of empty monochromatic triangles?

This problem was solved by Aichholzer et al. [4], when they proved that any bichromatic point set with n elements has at least $\Omega(n^{5/4})$ monochromatic triangles. This bound was improved to $\Omega(n^{4/3})$ by Pach and Tóth [31]. In fact we believe that the following conjecture is true.

Conjecture 4 [4]. Any bichromatic point set has a quadratic number of empty monochromatic triangles.

Another open question posed in [16], which is widely believed to be true, is the following.

Conjecture 5 [16]. Any large enough bichromatic point set always has a monochromatic 4-hole.

In Fig. 13 we show a point set with 12 points that has no monochromatic 4-hole. If we allow a hole to be non-convex; i.e., a not necessarily convex polygon whose vertices belong to P and which contains no elements of P in its interior, then if P is a bichromatic point set with at least 5044 points, it always has a (not necessarily convex) 4-hole [5]. Recently it was proved that any large enough

Fig. 13. A point set with no monochromatic 4-hole. (Color figure online)

bicolored point set has a monochromatic convex quadrilateral with at most one point in its interior [19].

In [34] Urrutia proved that any sufficiently large 4-colored point set in $\mathbb{R}^3$ always has an empty monochromatic tetrahedron. In fact we believe that this is true for any k-colored point set in $\mathbb{R}^3$ that is large enough. The following conjecture by Urrutia would imply Conjecture 5.

Conjecture 6 [34]**.** Any large enough point set in general position in $\mathbb{R}^3$ has a tetrahedralization with a super-linear number of tetrahedra.

In [5], Aichholzer et al. posed the following problem.

Problem 4 [5]**.** Is it true that any sufficiently large k-colored point set in R^3 in general position contains a convex monochromatic polyhedron that is the union of two interior disjoint tetrahedra that share a face?

We believe that this is true for any k.

9 A Set of New Problems in Geometric Number Theory

All point sets P with n elements considered in this section will be labelled with the integers $\{1, \ldots, n\}$ such that no two points have the same label. A triangulation of P is a partition of its convex hull into interior disjoint triangles whose vertices are in P. The weight of a triangle in T is the sum of the labels of its vertices. A triangulation T of P is called *graceful* if any two triangulations of T have different weights; see Fig. 14. The following conjecture was posed by Urrutia at the Indonesia–Japan Joint Conference on Combinatorial Geometry and Graph Theory in 2003.

Conjecture 7. Any point set in convex position has a graceful triangulation.

Since then, several variants to the graceful triangulation problem have been studied. An *island* is a point set that is the intersection of P with a convex set S. The weight of an island is the sum of the labels of its elements. In Fig. 15 we show two islands of a point set with weights 19 and 17.

Clearly any point set has islands of weights $\{1, \ldots, 2n-1\}$. Since each point is an island of P, islands with weights $\{1, 2, \ldots, n\}$ exist. The pairs of points of P with weights i and n are islands of weight $n+i$, $i = 1, \ldots, (n-1)$. The following open problem by Sakai and Urrutia has been around for about ten years.

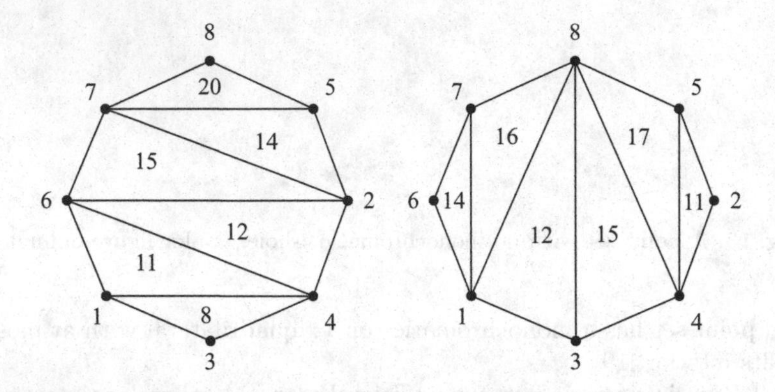

Fig. 14. Two graceful triangulations of a point set in convex position.

Conjecture 8. Any point set P in general position has an island of weight $2n$.

Another problem arising from the graceful triangulation problem is the following. A plane geometric graph with vertices in P is called *harmonic* if the weights of all of its edges (the sum of the labels of its endpoints) are different. In Araujo el al. [10], problems on the existence of harmonic paths and harmonic perfect matchings are studied. In particular, they study this question.

Question 1. What is the maximum size of a non-crossing harmonic matching (or path) that is always guaranteed in a configuration of N labelled points in general position in the plane?

They conjecture that any point set has a harmonic path (and thus matching) of length $\Theta(n)$.

They studied the following problem: Let $P = R \cup B$ such that $R \cup B$ is in convex position, $|R| = |B|$, such that R and B are separated by a line. Suppose that the points of R are labelled $0, 1, 2, \ldots, n-1$ in the cyclic order in which they appear on the convex hull of P, and that the labels of B form a permutation of $\{0, 1, 2, \ldots, n-1\}$.

Question 2. What is the number of edges in a maximum size harmonic alternating path (or matching) of P?

They proved the following.

Theorem 10 [10]. *If n is sufficiently large, then P has a simple harmonic alternating path of length at least $\frac{n^{2/3}}{9}$.*

Balogh, Pittel and Salazar [13] studied the same problem, but for point sets with n elements such that their elements have been been independently labelled at random with an integer in $\{0, 1, ..., n-1\}$, with each point receiving a unique label. Balogh et al. proved that, with high probability, there is a non-crossing (mod n)-harmonic matching covering at least $n/2 - (n \log n)^{1/3}$ points. Furthermore, with probability $(n^{-1/3} \log^{-1} n)$, there is a perfect non-crossing (mod n)-harmonic matching.

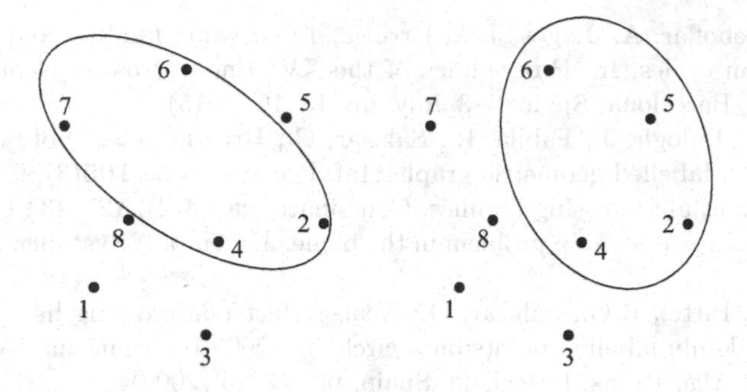

Fig. 15. Two islands of a point set with weights 19 and 17.

10 Final Remarks

In this paper we have surveyed some problems on geometric graphs on point sets. In some cases the points were colored, in others they were not. We also presented a number of open problems which we hope will attract the attention of some researchers. There are three recent surveys on similar problems that our readers may find interesting; these are Kano's survey "Discrete Geometry on Red and Blue Points in the Plane" [21], Abrego et al.'s [3] "SIGACT Computational Geometry Column" (2015), and Urrutia's paper [35].

Acknowledgment. We would like to thank the referees for their careful reports.

References

1. Abellanas, M., García, A., Hurtado, F., Tejel, J.: Caminos alternantes. In: Proceedings of the X Encuentros de Geometría Computacional: Sevilla, 16–17 junio 2003, pp. 7–12 (2003)
2. Abellanas, M., García, J., Hernández, G., Noy, M., Ramos, P.A.: Bipartite embeddings of trees in the plane. Discret. Appl. Math. **93**(2), 141–148 (1999)
3. Abrego, B., Dumitrescu, A., Fernández, S., Tóth, C.D.: Computational geometry column 61. ACM SIGACT News **46**(2), 65–77 (2015)
4. Aichholzer, O., Fabila-Monroy, R., Flores-Peñaloza, D., Hackl, T., Huemer, C., Urrutia, J.: Empty monochromatic triangles. Comput. Geom. **42**(9), 934–938 (2009)
5. Aichholzer, O., Hackl, T., Huemer, C., Hurtado, F., Vogtenhuber, B.: Large bichromatic point sets admit empty monochromatic 4-Gons. SIAM J. Discret. Math. **23**(4), 2147–2155 (2010)
6. Akiyama, J., Alon, N.: Disjoint simplices and geometric hypergraphs. Ann. New York Acad. Sci. **555**(1), 1–3 (1989)
7. Akiyama, J., Urrutia, J.: A note on balanced colourings for lattice points. Discret. Math. **83**(1), 123–126 (1990)
8. Akiyama, J., Urrutia, J.: Simple alternating path problem. Discret. Math. **84**(1), 101–103 (1990)

9. Luis, J., Rebollar, Á., Lagos, J.C., Urrutia, J.: Crossing families and self crossing Hamiltonian cycles. In: Proceedings of the XVI Encuentros de Geometría Computacional, Barcelona, Spain, 1–3 July, pp. 13–16 (2015)
10. Araujo, G., Balogh, J., Fabila, R., Salazar, G., Urrutia, J.: A note on harmonic subgraphs in labelled geometric graphs. Inf. Process. Lett. **105**(3), 98–102 (2008)
11. Aronov, B., et al.: Crossing families. Combinatorica **14**(2), 127–134 (1994)
12. Atallah, M.J.: A matching problem in the plane. J. Comput. Syst. Sci. **31**(1), 63–70 (1985)
13. Balogh, J., Pittel, B.G., Salazar, G.: Near-perfect non-crossing harmonic matchings in randomly labelled points on a circle. In: 2005 International Conference on Analysis of Algorithms, Barcelona, Spain, pp. 17–26 (2005)
14. Biedl, T.C., et al.: Balanced k-colorings. Discret. Math. **254**(1–3), 19–32 (2002)
15. Claverol, M., García, A., Garijo, D., Seara, C., Tejel, J.: On Hamiltonian alternating cycles and paths. Comput. Geom. **68**, 146–166 (2018)
16. Devillers, O., Hurtado, F., Károlyi, G., Seara, C.: Chromatic variants of the Erdos-Szekeres theorem on points in convex position. Comput. Geom. **26**(3), 193–208 (2003)
17. Erdős, P.: Some more problems on elementary geometry. Austral. Math. Soc. Gaz **5**(2), 52–54 (1978)
18. Erdős, P., Szekeres, G.: A combinatorial problem in geometry. Compositio Mathematica **2**, 463–470 (1935)
19. González-Martínez, A.C., Cravioto-Lagos, J., Urrutia, J.: Almost empty monochromatic polygons in planar point sets. In: Proceedings XVI Spanish Meeting on Computational Geometry, Barcelona, 1–3 de julio, 2015, pp. 81–84 (2015)
20. Horton, J.D.: Sets with no empty convex 7-Gons. Can. Math. Bull. **26**(4), 482–484 (1983)
21. Kaneko, A., Kano, M.: Discrete geometry on red and blue points in the plane - a survey. In: Aronov, B., Basu, S., Pach, J., Sharir, M. (eds.) Discrete and Computational Geometry: The Goodman-Pollack Festschrift, pp. 551–570. Springer, Heidelberg (2003). https://doi.org/10.1007/978-3-642-55566-4_25
22. Kaneko, A., Kano, M., Yoshimoto, K.: Alternating Hamilton cycles with minimum number of crossings in the plane. Int. J. Comput. Geom. Appl. **10**, 73–78 (2000)
23. Kano, M., Merino, C., Urrutia, J.: On plane spanning trees and cycles of multicolored point sets with few intersections. Inf. Process. Lett. **93**(6), 301–306 (2005)
24. Kynčl, J., Pach, J., Tóth, G.: Long alternating paths in bicolored point sets. Discret. Math. **308**(19), 4315–4321 (2008)
25. Leaños, J., Merino, C., Salazar, G., Urrutia, J.: Spanning trees of multicoloured point sets with few intersections. In: Akiyama, J., Baskoro, E.T., Kano, M. (eds.) IJCCGGT 2003. LNCS, vol. 3330, pp. 113–122. Springer, Heidelberg (2005). https://doi.org/10.1007/978-3-540-30540-8_13
26. Lo, C.-Y., Matoušek, J., Steiger, W.: Algorithms for ham-sandwich cuts. Discret. Comput. Geom. **11**(4), 433–452 (1994). https://doi.org/10.1007/BF02574017
27. Merino, C., Salazar, G., Urrutia, J.: On the intersection number of matchings and minimum weight perfect matchings of multicolored point sets. Graphs Comb. **21**(3), 333–341 (2005)
28. Merino, C., Salazar, G., Urrutia, J.: On the length of longest alternating paths for multicoloured point sets in convex position. Discret. Math. **306**(15), 1791–1797 (2006)
29. Pach, J., Rubin, N.N., Tardos, G.: Planar point sets determine many pairwise crossing segments. arXiv: 1904.08845 (2019)

30. Pach, J., Solymosi, J.: Halving lines and perfect cross-matchings. Contemp. Math. **223**, 245–250 (1999)
31. Pach, J., Tóth, G.: Monochromatic empty triangles in two-colored point sets. Discret. Appl. Math. **161**(9), 1259–1261 (2013)
32. Suzuki, K.: On the number of intersections of three monochromatic trees in the plane. In: Akiyama, J., Kano, M. (eds.) JCDCG 2002. LNCS, vol. 2866, pp. 261–272. Springer, Heidelberg (2003). https://doi.org/10.1007/978-3-540-44400-8_28
33. Tokunaga, S.: Intersection number of two connected geometric graphs. Inf. Process. Lett. **59**(6), 331–333 (1996)
34. Urrutia, J.: Coloraciones, tetraedralizaciones, y tetraedros vacíos en coloraciones de conjuntos de puntos en $\mathbb{R}^3$. In: Proceedings of the X Encuentros de Geometría Computacional: Sevilla, 16–17 junio 2003, pp. 95–100 (2003)
35. Urrutia, J.: The mathematics of Ferran Hurtado: a brief survey. In: Akiyama, J., Ito, H., Sakai, T. (eds.) JCDCGG 2015. LNCS, vol. 9943, pp. 277–292. Springer, Cham (2016). https://doi.org/10.1007/978-3-319-48532-4_25

The Two-Distance Sets in Dimension Four

Ferenc Szöllősi(✉)

Department of Communications and Networking, Aalto University School of Electrical Engineering, P.O. Box 15400, 00076 Aalto, Finland

Abstract. A finite set of vectors $\mathcal{X}$ in the d-dimensional Euclidean space $\mathbb{R}^d$ is called a 2-distance set, if the set of mutual distances between distinct elements of $\mathcal{X}$ has cardinality exactly 2. In this note we report, among other things, the results of a computer-aided enumeration of the 2-distance sets in $\mathbb{R}^4$.

1 Introduction

This paper is based on a talk given at the *21st Japan Conference on Discrete and Computational Geometry, Graphs, and Games* held in Manila, the Republic of the Philippines, September 1–3, 2018.

Let $d \geq 1$ be an integer, and let $\mathbb{R}^d$ denote the d-dimensional Euclidean space equipped with the standard inner product $\langle \cdot, \cdot \rangle$ and norm induced metric μ. Following the terminology of [14], a Euclidean representation of a simple graph Γ on $n \geq 1$ vertices is an embedding f (with real parameters $\alpha_2 > \alpha_1 > 0$) of the vertex set of Γ into $\mathbb{R}^d$, such that for different vertices $u \neq v$ we have $\mu(f(u), f(v)) = \alpha_1$ if and only if $\{u, v\}$ is an edge of Γ, and $\mu(f(u), f(v)) = \alpha_2$ otherwise. The smallest d for which such a representation exists is denoted by $\dim_2 \Gamma$. If Γ is neither complete, nor empty, then the set $\mathcal{X} := \{f(u) \colon u \text{ is a vertex of } \Gamma\}$ is called an n-element 2-distance set [2,5,15]. The representation f and its image $\mathcal{X}$ are called spherical, if $\mathcal{X}$ lies on the $(d-1)$-sphere of radius 1 in $\mathbb{R}^d$ [12,13]. A spherical representation is called J-spherical [12, Definition 4.1], if $\alpha_1 = \sqrt{2}$. Graphs on $n \geq d+2$ vertices having a J-spherical representation in $\mathbb{R}^d$ are in a certain sense extremal [18]. We remark that several authors follow a slightly different terminology, where the condition $\alpha_2 > \alpha_1$ is omitted, and therefore a given 2-distance set corresponds to a graph Γ and its complement $\overline{\Gamma}$ [11,16,19].

In this note we employ a computer-aided generation and classification of 2-distance sets in Euclidean spaces [21], a program initiated originally in [11]. In particular, we completely describe the 2-distance sets in $\mathbb{R}^4$ by determining all simple graphs Γ with $\dim_2 \Gamma = 4$. Since all such graphs are known on at most 6 vertices [5,14], and it is known that there are no such graphs on more than 10 vertices [11], the aim of this note is to close this gap by classifying the graphs (and in turn, the 2-distance sets) in the remaining cases. Our main result, which resolves a recent problem posed in [14, Section 4.3], is the following.

Theorem 1. *The number of n-element 2-distance sets in $\mathbb{R}^4$ for $n \in \{7, 8, 9\}$ is 33, 20, and 5 up to isometry.*

J. Akiyama et al. (Eds.): JCDCGGG 2018, LNCS 13034, pp. 18–27, 2021.
https://doi.org/10.1007/978-3-030-90048-9_2

It is known, see [10] (and also [15, Corollary 5] for an improvement), that if a 2-distance set in $\mathbb{R}^d$ has more than $2d+3$ elements, then necessarily $\alpha_1^2/\alpha_2^2 = (m-1)/m$ with an integer $m \geq 2$, $m \leq 1/2 + \sqrt{d/2}$. This property was leveraged earlier in [11] where the largest 2-distance sets in $\mathbb{R}^d$ for $d \leq 7$ were classified. The lack of sufficient understanding of the quantity α_1^2/α_2^2 is one reason why the results of Theorem 1 have not been obtained earlier.

The proof of Theorem 1 is in part computational, and follows from the theory developed earlier in [21], which we briefly outline here for completeness, as follows. Let a and b be indeterminates, and associate to an n-vertex graph Γ a "candidate Gram matrix" $G(a,b) := aA(\Gamma) + bA(\overline{\Gamma}) + I$, where $A(\Gamma)$ is the graph adjacency matrix, and I is the identity matrix of order n. We recall the following classical result.

Lemma 1. *Let Γ be an n-vertex graph with vertices $v_1, \ldots, v_n$, and let f be a spherical representation (with parameters $\alpha_2 > \alpha_1 > 0$ as usual) in $\mathbb{R}^d$. Then the Gram matrix of the representation can be written as*

$$[\langle f(v_i), f(v_j)\rangle]_{i,j=1}^n = G(1-\alpha_1^2/2, 1-\alpha_2^2/2), \tag{1}$$

which is a positive semi-definite matrix of rank at most d.

The correspondence given in Lemma 1 allows us to construct a spherical representation f, based solely on $A(\Gamma)$, by exploiting the fact that the rank of $G(1-\alpha_1^2/2, 1-\alpha_2^2/2)$ is at most d. Indeed, if we are given a candidate Gram matrix $G(a,b)$, then those values $a^*, b^* \in \mathbb{C}$ for which $G(a^*, b^*)$ has rank at most d can be found by considering the zero set of all $(d+1)\times(d+1)$ minors of $G(a,b)$. These minors are in fact polynomials in the variables a and b, and their zero set can be analyzed by a standard Gröbner basis computation [1,4], as detailed in [21] (see also [11, Section 7]). In particular, if these polynomial equations have no common solutions, then the candidate Gram matrix (as well as both Γ and its complement) cannot correspond to a spherical 2-distance set in $\mathbb{R}^d$. On the other hand, if some solutions are found, then we say that the candidate Gram matrix survived the test, and one should further ascertain that $G(a^*, b^*)$ is a positive semidefinite matrix. This can be done by, e.g., investigating the signs of the coefficients of its characteristic polynomial [7, Corollary 7.2.4]. Finally, the spherical 2-distance set $\mathcal{X}$ can be recovered (uniquely, up to isometry) from $G(a^*, b^*)$ through the Cholesky decomposition. The following example illustrates this approach.

Example 1. Consider the adjacency matrix of the complement of the cycle graph of length 7, Γ_{7O}, and the associated candidate Gram matrix given by:

$$A(\Gamma_{7O}) = \begin{bmatrix} 0\,1\,1\,1\,1\,0\,0 \\ 1\,0\,1\,1\,0\,1\,0 \\ 1\,1\,0\,0\,1\,0\,1 \\ 1\,1\,0\,0\,0\,1\,1 \\ 1\,0\,1\,0\,0\,1\,1 \\ 0\,1\,0\,1\,1\,0\,1 \\ 0\,0\,1\,1\,1\,1\,0 \end{bmatrix} \rightsquigarrow G(a,b) = \begin{bmatrix} 1\,a\,a\,a\,a\,b\,b \\ a\,1\,a\,a\,b\,a\,b \\ a\,a\,1\,b\,a\,b\,a \\ a\,a\,b\,1\,b\,a\,a \\ a\,b\,a\,b\,1\,a\,a \\ b\,a\,b\,a\,a\,1\,a \\ b\,b\,a\,a\,a\,a\,1 \end{bmatrix}.$$

First we show that Γ_{7O} has no spherical representation in $\mathbb{R}^3$. Suppose the contrary. Then there exist distinct real numbers $a^* \neq 1$ and $b^* \neq 1$, such that $rankG(a^*, b^*) \leq 3$, and in particular, all 4×4 minors of $G(a^*, b^*)$ are simultaneously 0. Let S_4 denote the set of polynomials formed by the 4×4 minors of $G(a, b)$, and consider the system of $1 + \binom{7}{4}^2$ polynomials $S_4 \cup \{u(a-1)(b-1)(a-b)+1\}$ in the variables u, a, and b. The role of the auxiliary variable u is to ascertain that the values taken by a and b are distinct, and different from 1. A routine Gröbner basis computation shows that the reduced Gröbner basis is $\{1\}$, meaning that there are no complex solutions to the previous system of equations. Since the role of a graph and its complement amounts to the change of variables $(a, b) \leftrightarrow (b, a)$ in the previous reasoning, it follows that neither Γ_{7O}, nor $\overline{\Gamma_{7O}}$ can be an induced subgraph of a graph having a spherical representation in $\mathbb{R}^3$. Of course, this result is well-known, as the largest cardinality of a 2-distance set in $\mathbb{R}^3$ is 6 (see [5, Theorem 8]).

Next we determine all spherical representations of Γ_{7O} in $\mathbb{R}^4$. As before, we consider the zero set of the $1 + \binom{7}{5}^2$ polynomials $S_5 \cup \{u(a-1)(b-1)(a-b)+1\}$ in the variables u, a, and b, where S_5 denotes the set of 5×5 minors of $G(a, b)$. In this case, a degree reverse lexicographic reduced Gröbner basis (with respect to some variable ordering) is

$$\{8a^3 + 32a^2 + 10a - 1, 2b + 4a + 1, 49u + 64a^2 + 240a + 24\}.$$

Since the discriminant of this cubic equation is positive, there are three real solutions ensuring $rankG(a^*, b^*) \leq 4$, which we give here approximately as follows: $(a^*, b^*) \in \{(-3.648, 6.796), (-0.431, 0.363), (0.079, -0.659)\}$. In each of these cases, the eigenvalues of $G(a^*, b^*)$ are 0 of multiplicity 3, and $(7 \pm \sqrt{112a^*(3a^* + 1) - 7})/4$, each of multiplicity 2. This shows that the first case $(a^*, b^*) \approx (-3.648, 6.796)$ actually leads to an indefinite matrix which has no geometric meaning in a Euclidean space. In the second case, we find that $\alpha_1 = \sqrt{2 - 2a^*} \approx 1.692$, and $\alpha_2 = \sqrt{2 - 2b^*} \approx 1.129$. Since $\alpha_1 > \alpha_2$, this is a spherical embedding of the cycle graph of length 7, $\overline{\Gamma_{7O}}$. Finally, in the third case we find $\alpha_1 = \sqrt{2 - 2a^*} \approx 1.357$, and $\alpha_2 = \sqrt{2 - 2b^*} \approx 1.821$. This solution corresponds to the unique spherical representation of Γ_{7O} in $\mathbb{R}^4$. The 2-distance set $\mathcal{X}$ can be (uniquely, up to isometry) reconstructed by applying the Cholesky decomposition on the matrix $G(a^*, b^*)$ obtained in the third case. □

The treatment of the general case (i.e., when f is not necessarily spherical) is analogous, but slightly more technical as the image of f should be translated to the origin first, and then the Gram matrix of this shifted set should be considered [11, Section 7.1], [21, Section 4].

Lemma 2. *Let Γ be an n-vertex graph with vertices $v_1, \ldots, v_n$, and let f be a Euclidean representation (with parameters $\alpha_2 > \alpha_1 > 0$ as usual) in $\mathbb{R}^d$. Then for every $i, j \in \{1, \ldots, n-1\}$, we have*

$$\begin{aligned}\langle f(v_i) - f(v_n),\ f(v_j) - f(v_n)\rangle &= \\ &\left(G(\alpha_1^2, \alpha_2^2)_{i,n} + G(\alpha_1^2, \alpha_2^2)_{j,n} - G(\alpha_1^2, \alpha_2^2)_{i,j} + I_{ij}\right)/2.\end{aligned} \quad (2)$$

Once again, (2) describes the entries of an $(n-1)\times(n-1)$ positive semidefinite matrix of rank at most d, which depends on $A(\Gamma)$ only. This rank condition can be treated in a similar way as discussed previously, see [21], and Example 2.

In [21] we have implemented an algorithm to generate candidate Gram matrices in a row-by-row fashion, keeping only those where the Gröbner basis computation reports some solutions. Recall that these are referred to as the surviving candidate Gram matrices. The main advantage of this approach over other avenues suggested in the literature [5,11,12] is that here the rather technical eigenvalue-analysis (ensuring positive semidefiniteness) can be postponed until the very end of the search. Indeed, the classification of the largest 6-distance sets in $\mathbb{R}^2$, the largest 4-distance sets in $\mathbb{R}^3$, and the largest 3-distance sets in $\mathbb{R}^4$ was carried out in this fashion [21].

2 The Two-Distance Sets in $\mathbb{R}^4$

In this section we report the results of the classification of all 2-distance sets in $\mathbb{R}^4$. A similar study was reported earlier in [5], where all 2-distance sets in $\mathbb{R}^2$ and $\mathbb{R}^3$ were accounted for.

In Table 1 we summarize the number of surviving candidate Gram matrices found by our backtrack search, and the number of corresponding 2-distance sets. The entry marked by an asterisk indicates that 6 out of the 42 cases are actually the maximum 2-distance sets in $\mathbb{R}^3$, see [5, Section 10]. The proof of Theorem 1 can be obtained by analyzing one by one the surviving candidate Gram matrices and the corresponding graphs on $n\in\{7,8,9\}$ vertices in a similar fashion as shown in Example 1. This can be conveniently done by computer algebra.

Table 1. The number of candidate Gram matrices and 2-distance sets in $\mathbb{R}^4$

n	6	7	8	9	10	11
#Graphs up to complements	78	522	6178	137352	6002584	509498932
#Surviving cand. Gram mat. (all)	77	22	13	4	1	0
#Surviving cand. Gram mat. (spherical)	30	17	6	2	1	0
#Spherical 2-distance sets	*42	23	7	2	1	0
#Nonspherical 2-distance sets	103	10	13	3	0	0

It is known that the maximum cardinality of a 2-distance set in $\mathbb{R}^4$ is exactly 10, and the unique configuration realizing this corresponds to the triangular graph $T(5)$, see [11]. We have verified this result independently. Indeed, our computer program identified a single 10×10 candidate Gram matrix, which cannot be extended any further, and whose spherical representation is shown in Table 2. While the subgraphs of $T(5)$ obviously correspond to various spherical 2-distance sets embedded in $\mathbb{R}^4$, there are several additional examples as follows.

Proposition 1 (cf. [11, Section 6.2]). *The number of* 9*-point* 2*-distance sets in* $\mathbb{R}^4$ *is* 5*, out of which* 2 *are spherical.*

Proof. Our computer program generated 4 candidate Gram matrices in the general case, and 2 in the spherical case, see Table 1. The two spherical cases correspond to the 9-vertex subgraph of $T(5)$, and to the Paley graph, see Table 2. The remaining two candidate Gram matrices correspond to the three nonspherical 2-distance sets with $\alpha_1^2/\alpha_2^2 = (3-\sqrt{5})/2$ (see Table 6) discovered earlier in [11].

Proposition 2. *The number of* 8*-point* 2*-distance sets in* $\mathbb{R}^4$ *is* 20*, out of which* 7 *are spherical.*

Proof. Our computer program generated 13 candidate Gram matrices in the general case, and 6 in the spherical case, see Table 1. In the spherical case five out of the six candidate Gram matrices yielded one spherical 2-distance set, while one resulted in two, see Table 2. The remaining 7 candidate Gram matrices correspond to two nonspherical 2-distance sets each, except for the one whose underlying graph is self-complementary. See Table 6 for some details.

Proposition 3. *The number of* 7*-point* 2*-distance sets in* $\mathbb{R}^4$ *is* 33*, out of which* 23 *are spherical.*

Proof. Our computer program generated 22 candidate Gram matrices in the general case, and 17 in the spherical case, see Table 1. In the spherical case there was a single matrix which did not correspond to any 2-distance sets as it turned out to be indefinite. All the other candidate Gram matrices yielded at least one spherical 2-distance set, see Table 3. The remaining 5 candidate Gram matrices correspond to two nonspherical 2-distance sets each, see Table 6.

One, perhaps surprising, consequence of the classification results highlighted in Proposition 1–3 is that $\alpha_1^2/\alpha_2^2 \in \{1/2, (3-\sqrt{5})/2\}$ for the n-element 2-distance sets in $\mathbb{R}^4$ with $n \geq 7$, except for the cycle graph of length 7 and its complement (see entry Γ_{7O} in Table 3, and Example 1). It would be interesting to find a theoretical reason for this phenomenon (cf. [10]).

Proposition 4 (cf. [5, p. 494], [14, Section 4.3]). *The number of* 6*-point* 2*-distance sets in* $\mathbb{R}^4$ *is* 145*. The number of* 6*-point spherical* 2*-distance sets in* $\mathbb{R}^4$ *is* 42*, out of which* 6 *are in fact the maximum* 2*-distance sets in* $\mathbb{R}^3$*.*

Proof. It is known, see [5,14], that a graph on 6 vertices can be represented in $\mathbb{R}^4$ unless it is a disjoint union of cliques. Since the total number of simple graphs on 6 vertices is 156, out of which 11 are disjoint union of cliques, we find that 145 graphs can be represented in $\mathbb{R}^4$. Our computer program generated 30 candidate Gram matrices in the spherical case, see Table 1. There were two indefinite matrices, and the remaining 28 resulted in at least one spherical 2-distance set each. Amongst these, we found the 6 maximum 2-distance sets in $\mathbb{R}^3$, denoted by Γ_{6K}, Γ_{6O}, $\overline{\Gamma}_{6O}$, Γ_{6R}, $\overline{\Gamma}_{6R}$, and Γ_{6Y}, see Table 4.

Finally, there are 7 graphs Γ on $n = 5$ vertices for which $\dim_2 \Gamma = 4$. These are precisely the graphs formed by disjoint union of cliques [5,14]. One particular spherical representation of these is given in Table 5. The number of corresponding nonisometric 2-distance sets in $\mathbb{R}^4$ in these cases is infinite.

Corollary 1. *The number of graphs Γ for which $\dim_2 \Gamma = 4$ is* 205.

Proof. This follows from earlier results in [5,11], and Theorem 1: the number of such graphs on $n \in \{5, 6, 7, 8, 9, 10\}$ vertices is 7, 139, 33, 20, 5, and 1, respectively, and there are no such graphs on $n < 5$ or $n > 10$ vertices.

It would be interesting to find a combinatorial characterization of these graphs.

3 On Distances Occurring a Different Number of Times

In this section we report on some results regarding 8-element 7-distance sets in $\mathbb{R}^2$. Sets with "few" distances were investigated earlier in [3,6,9,20].

An old problem of ERDŐS asks if it is possible to have n points in general position in the plane (no three on a line or four on a circle) such that there is a distance determined by the points that occurs exactly i times for every $i \in \{1, \ldots, n-1\}$ (see [17] and the references therein). Sets up to 8 points are known, but the cases $n \geq 9$ are open. Our computational approach can address this problem too, and we report the following results of an incomplete search.

Example 2. Here we construct two nonisometric sets of $n = 8$ points in $\mathbb{R}^2$ with the property that there are distances occurring once, twice, ..., seven times between the points. Consider the following candidate Gram matrix, in which entry a appears twice, entry b appears four times, ..., entry g appears 14 times:

$$G(a,b,c,d,e,f,g) := \begin{bmatrix} 1 & a & b & e & e & e & f & g \\ a & 1 & g & b & f & g & f & e \\ b & g & 1 & f & c & g & g & f \\ e & b & f & 1 & g & e & d & f \\ e & f & c & g & 1 & d & c & d \\ e & g & g & e & d & 1 & g & d \\ f & f & g & d & c & g & 1 & c \\ g & e & f & f & d & d & c & 1 \end{bmatrix}.$$

As outlined in Sect. 1, we are interested in determining the real values $a^*, \ldots, g^*$ for which the matrix $[M]_{i,j=1}^{7} := (G_{i,8} + G_{j,8} - G_{i,j} + I_{i,j})/2$ is positive semidefinite of rank at most 2. Up to global isometry, we may assume that $g^* = 1$. Now let S_3 denote the set of 3×3 minors of M, and consider the set of polynomials

$$S_3 \cup \{u(a-b)(a-c)\cdots(a-f)\cdots(e-f)\cdots a(a-1)\cdots f(f-1)+1\}.$$

The polynomial featuring the auxiliary variable u ensures that the unknown distances are pairwise distinct, and each is distinct from 0 and 1 (which is the already assigned value of g). In all, this is a system of $1 + \binom{7}{3}^2$ polynomials in the 7 unknown variables $u, a, \ldots, f$. We compute a degree reverse lexicographic reduced Gröbner basis, which is $\{a^2 - 12a + 12, 2b - a - 4, 2c + a - 12, d - 2, 2e - a, f - 3, 106375680000u + 274469a - 2991432\}$. Therefore, we have two sets of solutions: $(a^*, b^*, \ldots, g^*) \in \{(6+2\sqrt{6}, 5+\sqrt{6}, 3-\sqrt{6}, 2, 3+\sqrt{6}, 3, 1), (6-2\sqrt{6}, 5-\sqrt{6}, 3+\sqrt{6}, 2, 3-\sqrt{6}, 3, 1)\}$. Applying the Cholesky decomposition to either of the positive semidefinite matrices $M(a^*, \ldots, g^*)$ yields a set of 7 points on the plane. Together with the origin these form a set of 8 points in general position, with distances $\alpha_1 = \sqrt{a^*}, \ldots, \alpha_7 = \sqrt{g^*}$. □

Theorem 2. *There exist at least* 5 *pairwise nonisometric planar configurations of* $n = 8$ *points in general position, where there is a distance determined by the points that occurs exactly* i *times for every* $i \in \{1, \ldots, 7\}$.

Proof. See Example 2, and the papers [8,17].

We believe that an exhaustive search for the case $n = 9$ is soon within reach.

Acknowledgements. This research was supported in part by the Academy of Finland, Grant #289002.

A Tables of Data

In the tables the vectorization (i.e., row-wise concatenation) of the lower triangular part of a graph adjacency matrix of order n is denoted by a string of letters a and b of length $n(n-1)/2$, where letter a indicates adjacent vertices. The ordered pair (a^*, b^*) indicates the values for which the matrix $G(a^*, b^*)$ is positive semidefinite.

Table 2. Spherical 2-distance sets on $n \in \{8, 9, 10\}$ points in $\mathbb{R}^4$

n	$G(a,b)$	$G(1,0)$	(a^*, b^*)	Remark
10	aaaaaaaabbababaabbaaabaabaabbabaabaabbaabaaaa	Γ_{10A}	$(1/6, -2/3)$	$T(5)$, $\dim_2(\overline{\Gamma}_{10A}) = 5$
9	aaaaaaaabbababaabbaaabaabaabbabaabaa	Γ_{9A}	$(1/6, -2/3)$	$\Gamma_{9A} \sim \Gamma_{10A} \setminus \{*\}$, $\dim_2(\overline{\Gamma}_{9A}) = 5$
9	aaaabbabbabababbabbaabbaababbbababaa	Γ_{9B}	$(1/4, -1/2)$	Paley, self-complementary
8	aaaaaaaaabaabaaabaaaabaaaaaa	Γ_{8A}	$(0, -1)$	16-cell, J-spherical, $\dim_2(\overline{\Gamma}_{8A}) = 7$
8	aaaaaaaaabaabababaabbbaaabbb	Γ_{8B}	$((1 \pm \sqrt{5})/4, 1/2)$	
8	aaaaaaaabbababababaabbbaabaa	Γ_{8C}	$(1/6, -2/3)$	$\dim_2(\overline{\Gamma}_{8C}) = 5$
8	aaaaaaabbbbabbabbabaabbbaaaa	Γ_{8D}	$(1/5, -3/5)$	$\dim_2(\overline{\Gamma}_{8D}) = 6$
8	aaaaaaaabbababaabbaaabaabaab	Γ_{8E}	$(1/6, -2/3)$	$\Gamma_{8E} \sim \Gamma_{9A} \setminus \{*\}$, $\dim_2(\overline{\Gamma}_{8E}) = 5$
8	aaaabbabbabababbabbaabbaabab	Γ_{8F}	$(1/4, -1/2)$	$\Gamma_{8F} \sim \Gamma_{9B} \setminus \{*\}$, self-complementary

Table 3. Spherical 2-distance sets on $n = 7$ points in $\mathbb{R}^4$

n	$G(a,b)$	$G(1,0)$	(a^*,b^*)	Remark
7	aaaaaaaaabaabaaabaaaa	Γ_{7A}	$(0,-1)$	J-spherical, $\dim_2(\overline{\Gamma}_{7A}) = 6$
7	aaaaaaaaabaababaabbaa	Γ_{7B}	$((1 \pm \sqrt{5})/4, 1/2)$	
7	aaaaaaaaabaabababaabb	Γ_{7C}	$((1 \pm \sqrt{5})/4, 1/2)$	
7	aaaaaaaabbababaabbaaa	Γ_{7D}	$(1/6, -2/3)$	$\dim_2(\overline{\Gamma}_{7D}) = 5$
7	aaaaaaaabbabababababaab	Γ_{7E}	$(1/6, -2/3)$	$\dim_2(\overline{\Gamma}_{7E}) = 5$
7	aaaaaaaabbababbabbabb	Γ_{7F}	$((1 \pm \sqrt{5})/4, 1/2)$	
7	aaaaaaaabbababbbaabbb	Γ_{7G}	$((1 \pm \sqrt{5})/4, 1/2)$	
7	aaaaaaabbbbabbabbabaa	Γ_{7H}	$(1/5, -3/5)$	$\dim_2(\overline{\Gamma}_{7F}) = 5$
7	aaaaaaaabbababababaabaa	Γ_{7I}	$(1/6, -2/3)$	$\dim_2(\overline{\Gamma}_{7G}) = 5$
7	aaaaababaabaaaabbbbbb	Γ_{7J}	$(1/4, -1/2)$	$\dim_2(\overline{\Gamma}_{7H}) = 5$
7	aaaaababaabaaabbbaaaa	Γ_{7K}	$((-1-\sqrt{5})/8, 3(-1+\sqrt{5})/8)$	$\dim_2(\Gamma_{7K}) = 5$
7	aaaaababaabaabbbbbaaa	Γ_{7L}	$(1/6, -2/3)$	$\dim_2(\overline{\Gamma}_{7L}) = 5$
7	aaaaabababababbaaabbbbb	Γ_{7M}	$((1 \pm \sqrt{5})/4, 1/2)$	
7	aaaaabababababbababbbaa	Γ_{7N}	$(-5/12, 7/24)$	$\dim_2(\Gamma_{7N}) = 5$
7	aaaaababababbabaabbaaaa	Γ_{7O}	$(\alpha_\pm, -2\alpha_\pm - 1/2)$	$8\alpha^3 + 32\alpha^2 + 10\alpha - 1 = 0, \lvert\alpha_\pm\rvert \le 1$
7	aaaabbabbabababbabbaa	Γ_{7P}	$((1 \pm 3)/8, (-1 \mp 3)/8)$	

Table 4. Spherical 2-distance sets on $n = 6$ points in $\mathbb{R}^4$

n	$G(a,b)$	$G(1,0)$	(a^*,b^*)	Remark
6	aaaaaaaaabaabaa	Γ_{6A}	$(0,-1)$	J-spherical, $\dim_2(\overline{\Gamma}_{6A}) = 5$
6	aaaaaaaaabaabab	Γ_{6B}	$((1 \pm \sqrt{5})/4, 1/2)$	
6	aaaaaaaabbaabba	Γ_{6C}	$(-1/3, 1/3)$	$\dim_2(\Gamma_{6C}) = 4$
6	aaaaaaaabbababa	Γ_{6D}	$(1/6, -2/3)$	$\dim_2(\overline{\Gamma}_{6D}) = 4$
6	aaaaaaaabbababb	Γ_{6E}	$((1 \pm \sqrt{5})/4, 1/2)$	
6	aaaaaaaabbbbaaa	Γ_{6F}	$(0, \pm\sqrt{2}/2)$	Γ_{6F} is J-spherical
6	aaaaaaaabbbbaab	Γ_{6G}	$((-1 \pm 3)/12, (-1 \mp 3)/6)$	
6	aaaaaaabbbbabba	Γ_{6H}	$(1/5, -3/5)$	$\dim_2(\overline{\Gamma}_{6H}) = 4$
6	aaaaaaabbbbabbb	Γ_{6I}	$(1/2, (1 \pm \sqrt{5})/4)$	
6	aaaaababaaabbbb	Γ_{6J}	$(1/3, -1/3)$	$\dim_2(\overline{\Gamma}_{6J}) = 4$
6	aaaaababaabaaaa	Γ_{6K}	$(0,-1)$	octahedron, J-spherical, $\dim_2(\overline{\Gamma}_{6K}) = 5$
6	aaaaababaabaaab	Γ_{6L}	$((-1-\sqrt{5})/8, 3(-1+\sqrt{5})/8)$	$\dim_2(\Gamma_{6L}) = 4$
6	aaaaababaabaabb	Γ_{6M}	$(1/6, -2/3)$	$\dim_2(\overline{\Gamma}_{6M}) = 4$
6	aaaaababaabbbbb	Γ_{6N}	$(1/4, -1/2)$	$\dim_2(\overline{\Gamma}_{6N}) = 4$
6	aaaaabababababbaa	Γ_{6O}	$(\pm 1/\sqrt{5}, \mp 1/\sqrt{5})$	pentagonal pyramids [5, Figure 4]
6	aaaaabababbab	Γ_{6P}	$(-5/12, 7/24)$	$\dim_2(\Gamma_{6P}) = 4$
6	aaaaabababbbb	Γ_{6Q}	$((1 \pm \sqrt{5})/4, 1/2)$	
6	aaaaabababbaabb	Γ_{6R}	$(\pm 1/\sqrt{5}, \mp 1/\sqrt{5})$	pentahedrons [5, Figure 4]
6	aaaaabababbabaa	Γ_{6S}	$(\alpha_\pm, -2\alpha_\pm - 1/2)$	$8\alpha^3 + 32\alpha^2 + 10\alpha - 1 = 0, \lvert\alpha_\pm\rvert \le 1$
6	aaaaabababbabab	Γ_{6T}	$(\beta_\pm, 2\beta_\pm^2 - \beta_\pm - 1/2)$	$8\beta^3 - 8\beta^2 - 2\beta + 1 = 0, \lvert\beta_\pm\rvert \le 1$
6	aaaaabababbbbaa	Γ_{6U}	$(1/6, -2/3)$	$\dim_2(\overline{\Gamma}_{6U}) = 4$
6	aaaaababbbbabba	Γ_{6V}	$(1/3, -1/3)$	$\dim_2(\overline{\Gamma}_{6V}) = 4$
6	aaaaabbbaabbaaa	Γ_{6W}	$(-3/7, 2/7)$	$\dim_2(\Gamma_{6W}) = 4$
6	aaaabbabbababab	Γ_{6X}	$((-1 \pm 3)/8, (-1 \mp 3)/8)$	
6	aaaabbbababbaaa	Γ_{6Y}	$((-5 \pm 9)/28, (-13 \mp 27)/56)$	Γ_{6Y} is a square-faced triangular prism [5, Figure 2]
6	aaaabbbababbbaa	Γ_{6Z}	$(1/5, -3/5), (-\sqrt{2}/3, (\sqrt{2}-1)/3)$	
6	aaaabbbbbabbbaa	$\Gamma_{6Æ}$	$((-1 \pm 3\sqrt{3})/13, (-7 \mp 5\sqrt{3})/26)$	
6	aaabbbbbbabbbaa	$\Gamma_{6Œ}$	$(-1/2, 0)$	J-spherical, $\dim_2(\Gamma_{6Œ}) = 5$

Table 5. Examples of spherical 2-distance sets on $n = 5$ points in $\mathbb{R}^4$

n	$G(a,b)$	$G(1,0)$	(a^*,b^*)	Remark
5	aaaaaaaaaa	Γ_{5A}	$a^* = -1/4$	regular 5-cell, $\dim_2(\overline{\Gamma}_{5A}) = 4$, 1-distance set
5	aaaaaaaaab	Γ_{5B}	$((1-\sqrt{7})/6, 0)$	J-spherical, $\dim_2(\Gamma_{5B}) = 3$
5	aaaaaabbbb	Γ_{5C}	$(0, -1/2)$	J-spherical, $\dim_2(\overline{\Gamma}_{5C}) = 3$
5	aaaaabaabb	Γ_{5D}	$(-1/3, 0)$	J-spherical, $\dim_2(\Gamma_{5D}) = 3$
5	aaaaababaa	Γ_{5E}	$((1-\sqrt{5})/4, 0)$	J-spherical, $\dim_2(\Gamma_{5E}) = 3$
5	aaabbbbbba	Γ_{5F}	$(0, -1/\sqrt{6})$	J-spherical, $\dim_2(\overline{\Gamma}_{5F}) = 3$

Table 6. General (nonspherical) 2-distance sets on $n \in \{7, 8, 9\}$ points in $\mathbb{R}^4$

n	$G(a, b)$	$G(1, 0)$	(a^*, b^*)	Remark
9	aaaaaaaaaaaabbbababbbabbabbbabbbabbb	Γ_{9C}	$(1, (3 \pm \sqrt{5})/2)$	truncated pyramids [11]
9	aaaaaaaaabaababaabbaaabbbbbbbabbbbbb	Γ_{9D}	$(1, (3 + \sqrt{5})/2)$	self-complementary, [11]
8	aaaaaaaaaaaabbbababbbabbabbb	Γ_{8G}	$(1, (3 \pm \sqrt{5})/2)$	$\Gamma_{8G} \sim \Gamma_{9C} \setminus \{*\}$
8	aaaaaaaaaaaabbbababbbbaabbbb	Γ_{8H}	$(1, (3 \pm \sqrt{5})/2)$	
8	aaaaaaaaabaababaabbaaabbbbbb	Γ_{8I}	$(1, (3 \pm \sqrt{5})/2)$	$\Gamma_{8I} \sim \Gamma_{9D} \setminus \{*\}$
8	aaaaaaaaabaababaabbaabbbbbbb	Γ_{8J}	$(1, (3 \pm \sqrt{5})/2)$	
8	aaaaaaaaabaabababaabbabbbbbb	Γ_{8K}	$(1, (3 \pm \sqrt{5})/2)$	
8	aaaaaaaaabaabababaabbbbbbbbb	Γ_{8L}	$(1, (3 \pm \sqrt{5})/2)$	$\Gamma_{8L} \sim \overline{\Gamma}_{9C} \setminus \{*\}$
8	aaaaaaaaabaabababbbbbbabbbbb	Γ_{8M}	$(1, (3 + \sqrt{5})/2)$	$\Gamma_{8M} \sim \Gamma_{9D} \setminus \{*\}$, self-complementary
7	aaaaaaaaaaaabbbababbb	Γ_{7Q}	$(1, (3 \pm \sqrt{5})/2)$	$\Gamma_{7Q} \sim \Gamma_{9C} \setminus \{*, *\}$
7	aaaaaaaaabaabababbbbb	Γ_{7R}	$(1, (3 \pm \sqrt{5})/2)$	$\Gamma_{7R} \sim \Gamma_{9D} \setminus \{*, *\}$
7	aaaaaaaaabaababbbbbbb	Γ_{7S}	$(1, (3 \pm \sqrt{5})/2)$	$\Gamma_{7S} \sim \overline{\Gamma}_{9C} \setminus \{*, *\}$
7	aaaaaaaaababbbbbabbbb	Γ_{7T}	$(1, (3 \pm \sqrt{5})/2)$	$\Gamma_{7T} \sim \Gamma_{9D} \setminus \{*, *\}$
7	aaaaaaaabbababbabbbbb	Γ_{7U}	$(1, (3 \pm \sqrt{5})/2)$	$\Gamma_{7U} \sim \Gamma_{9C} \setminus \{*, *\}$

References

1. Abbott, J., Bigatti, A.M.: CoCoALib: a C++ library for doing Computations in Commutative Algebra (2018). http://cocoa.dima.unige.it/cocoalib, ver. 0.99560
2. Bannai, E., Bannai, E., Stanton, D.: An upper bound for the cardinality of an s-distance subset in real Euclidean space, II. Combinatorica **3**, 147–152 (1983)
3. Bannai, E., Sato, T., Shigezumi, J.: Maximal m-distance sets containing the representation of the Johnson graph $J(n, m)$. Discret. Math. **312**, 3283–3292 (2012)
4. Becker, T., Weispfenning, V.: Gröbner Bases. Springer, New York (1993)
5. Einhorn, S.J., Schoenberg, I.J.: On Euclidean sets having only two distances between points I-II. Indag. Math. **69**, 479–504 (1966)
6. Erdős, P., Fishburn, P.: Maximum planar sets that determine k distances. Discret. Math. **160**, 115–125 (1996)
7. Horn, R.A., Johnson, C.R.: Matrix Analysis, 2nd edn. Cambridge University Press, Cambridge (2013)
8. Jafari, A., Amin, A.N.: On the Erdős distance conjecture in geometry. Open J. Discret. Math. **6**, 109–160 (2016)
9. Lan, W., Wei, X.: Classification of seven-point four-distance sets in the plane. Math. Notes **93**, 510–522 (2013)
10. Larman, D.G., Rogers, C.A., Seidel, J.J.: On two-distance sets in Euclidean space. Bull. Lond. Math. Soc. **9**(3), 261–267 (1977)
11. Lisoněk, P.: New maximal two-distance sets. J. Combin. Theory Ser. A **77**, 318–338 (1997)
12. Musin, O.R.: Graphs and spherical two-distance sets. Eur. J. Combin. **80**, 311–325 (2019)
13. Musin, O.R.: Spherical two-distance sets. J. Combin. Theory Ser. A **116**, 988–995 (2009)
14. Musin, O.R.: Towards a proof of the 24-cell conjecture. Acta Math. Hungar. **155**(1), 184–199 (2018). https://doi.org/10.1007/s10474-018-0828-5
15. Neumaier, A.: Distance matrices, dimension, and conference graphs. Indag. Math. **84**, 385–391 (1981)
16. Nozaki, H., Shinohara, M.: A geometrical characterization of strongly regular graphs. Linear Algebra Appl. **437**, 2587–2600 (2012)
17. Palásti, I.: Lattice-point examples for a question of Erdős. Period. Math. Hungar. **20**, 231–235 (1989)

18. Rankin, R.A.: The closest packing of spherical caps in n dimensions. Glasg. Math. J. **2**, 139–144 (1955)
19. Roy, A.: Minimal Euclidean representation of graphs. Discret. Math. **310**, 727–733 (2010)
20. Shinohara, M.: Classification of three-distance sets in two dimensional Euclidean space. Eur. J. Combin. **25**, 1039–1058 (2004)
21. Szöllősi, F., Östergård, P.R.J.: Constructions of maximum few-distance sets in Euclidean spaces. Electron. J. Combin. **27** #P1.23 (2020)

Negative Instance for the Edge Patrolling Beacon Problem

Zachary Abel[1], Hugo A. Akitaya[2], Erik D. Demaine[1], Martin L. Demaine[1], Adam Hesterberg[1], Matias Korman[3], Jason S. Ku[1], and Jayson Lynch[1](✉)

[1] Massachusetts Institute of Technology, Cambridge, USA
{zabel,edemaine,mdemaine,achester,jasonku,jaysonl}@mit.edu
[2] Carleton University, Ottawa, Canada
[3] Tufts University, Boston, USA
matias.korman@tufts.edu

Abstract. Can an infinite-strength magnetic beacon always "catch" an iron ball, when the beacon is a point required to be remain non-strictly outside a polygon, and the ball is a point always moving instantaneously and maximally toward the beacon subject to staying non-strictly within the same polygon? Kouhestani and Rappaport [JCDCG 2017] gave an algorithm for determining whether a ball-capturing beacon strategy exists, while conjecturing that such a strategy always exists. We disprove this conjecture by constructing orthogonal and general-position polygons in which the ball and the beacon can never be united.

Keywords: Beacon routing · Edge patrolling · Counterexample

1 Introduction

Suppose you have a metallic iron ***ball***, modeled as a point constrained to be within some compact polygonal region P. To manipulate the ball's location, you have one or more strongly magnetic ***beacons***, also modeled as points. When a beacon is ***active***, the ball instantaneously moves maximally toward the beacon, subject to staying within P; refer to Fig. 2. More precisely, the ball moves toward the board in a straight line until it either reaches the beacon or hits the polygon's boundary ∂P. In the latter case, the ball continues gliding along ∂P as long as the distance to the beacon decreases monotonically; the ball may later leave the boundary ∂P and resume moving along a straight line to the beacon. The ball stops moving either when reaching the beacon or on a boundary edge perpendicular to the segment connecting the ball and beacon. The general goal is to "capture" the ball by moving it to one of the beacons.

Biro et al. [2,3] introduced this model as a generalization of the art gallery problem, inspired by geographical greedy routing in sensor networks. In their problem, beacons are stationary (cannot move) and are only active when toggled into that state (with at most one beacon active at any time). The goal is to place the fewest possible beacons within P to enable routing the ball from a given start

J. Akiyama et al. (Eds.): JCDCGGG 2018, LNCS 13034, pp. 28–35, 2021.
https://doi.org/10.1007/978-3-030-90048-9_3

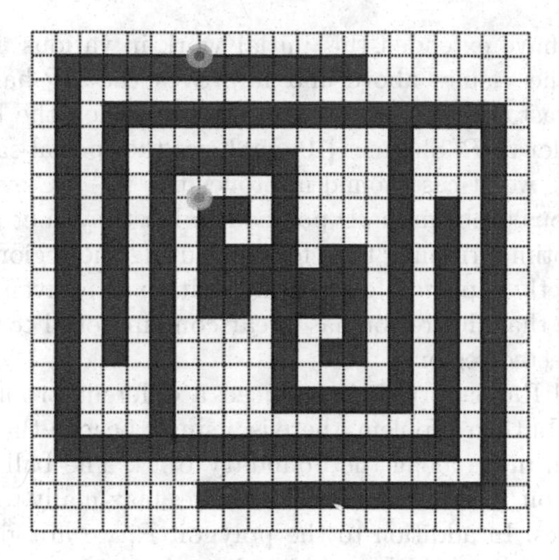

Fig. 1. Orthogonal counterexample to Kouhestani and Rappaport's conjecture [9]. The initial ball location is the grey dot; the initial beacon location is the orange dot; and the polygon is the purple region with black outline. (Color figure online)

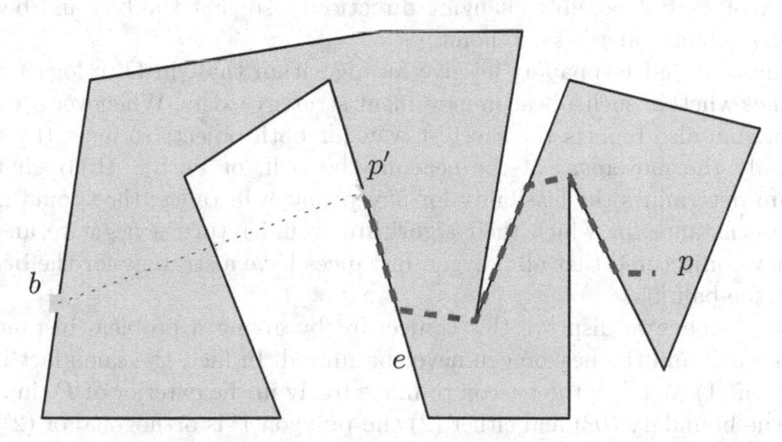

Fig. 2. The beacon attraction model of [2,3,8]. When the beacon b is active, it attracts the ball p along the dashed red trajectory. The ball p moves within the interior and boundary of P while greedily reducing its Euclidean distance to b. Its movement stops when the ball reaches a point p' (shown by a cross) on some edge e of P for which no local movement can reduce its distance to b. In particular, the segment $\overline{p'b}$ and the edge e of P form a right angle. (Color figure online)

point s to a given destination point t (which will naturally have a beacon) by activating beacons one at a time. Similar to the classic art gallery problem, they proved that this problem is NP-hard to solve exactly, and proved that $\lfloor n/2 \rfloor - 1$ beacons are always sufficient and sometimes necessary, where n is the number of vertices of a simple polygon P.

Several papers have extended this initial work in various directions. Biro's thesis [1] details the results above and improves the NP-hardness to APX-hardness. By contrast, Biro et al. [4] give a PTAS when the beacons are also allowed to be exterior to P. Shermer [10] analyzes the special case of orthogonal polygons, where the worst-case bound improves to $\lfloor \frac{n-4}{3} \rfloor$. Cleve and Mulzer [5] analyze the analogous problem in 3D polyhedra. Kostitsyna et al. [6] prove that $\Theta(n \log n)$ is the optimal running time for computing the region of beacon positions that can directly capture a given ball position in a given simple polygon. Biro et al. [4] prove that this region has linear combinatorial complexity but can have $\Omega(n)$ disconnected regions.

Kouhestani and Rappaport [8] introduced a different problem which is the focus of our paper. In their problem, there is a single beacon that is permanently active and that can move along the boundary of P. The ball moves infinitely faster than the beacon, so it instantaneously moves maximally toward the beacon as the beacon moves. In addition to the polygon P, we are given the starting location of both the ball p (which could be either in the interior or on boundary of P) and the beacon b (which can only be on the boundary). The goal is to give a beacon movement strategy, which moves the beacon continuously along the boundary of P (but possibly changing directions), so that the ball and beacon eventually coincide at the same point.

Kouhestani and Rappaport [8] give an algorithm that, in $O(n^3 \log n)$ time, determines whether such a beacon movement strategy exists. Whenever it exists, the algorithm also reports the shortest way for both objects to meet (by measuring only the movement of the beacon, the ball, or both). Although their algorithm determines the feasibility for any problem instance, they could never design an instance for which their algorithm would return a negative answer. Thus they conjectured that all polygon instances have a strategy for the beacon to reach the ball [9].

In this paper, we disprove this conjecture by giving a problem instance in which the ball and the beacon can never be united. In fact, the same fact holds true even if (1) we allow the beacon to move freely in the exterior of P (in addition to the boundary ∂P) and either (2) the polygon P is orthogonal or (2′) the polygon P is in general position. Figure 3 shows the orthogonal counterexample; the general-position example is a small perturbation thereof. An interactive version of Fig. 3 and other examples (implemented in a grid-based discrete model) can be found at http://erikdemaine.org/attractor/.

Our example is tight in the sense that no more freedom to the movement of the beacon can be added: if we allow the beacon to move in the interior of P, then the ball can be easily captured by walking along the geodesic from the starting positions (a fact observed in [7]).

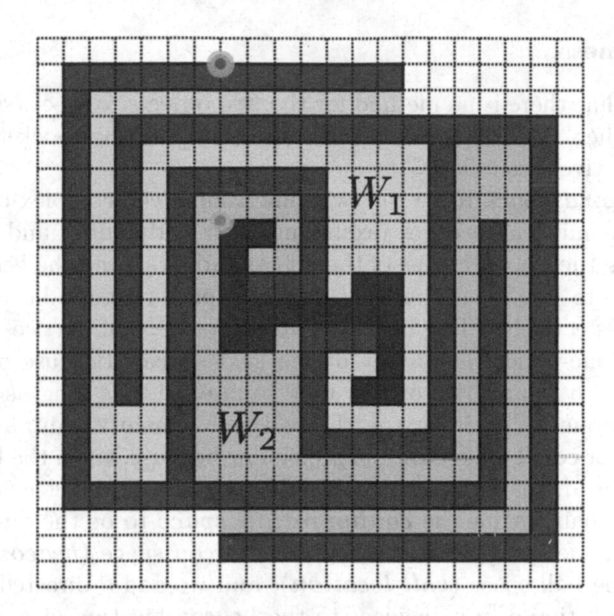

Fig. 3. Orthogonal counterexample to Kouhestani and Rappaport's conjecture [9] from Fig. 1 with added details. The initial ball location is the grey dot; the initial beacon location is the orange dot; and the polygon is the (light and dark) purple region with black outline. The dark-purple region is called the ***core***, and the two orange/pink regions W_1 and W_2 are called ***pockets***. (Color figure online)

2 Grid-Orthogonal Counterexample

2.1 Construction

Refer to Fig. 3. The key of the construction is the region shown in dark purple, called the ***core***, that will always contain the ball. If we partition the core in the middle by a vertical line, we obtain two rotationally symmetrical halves called ***hooks***. The union of the shaded regions forms the orthogonal convex hull of P. The two orange/pink regions denoted W_1 and W_2 are called ***pockets***.

The core has the following basic properties. If we place the ball and beacon at the initial positions shown in Fig. 3, and move the beacon clockwise along the boundary of the orthogonal convex hull of P, then the ball will always stay in the same hook. If we move the beacon counterclockwise along the boundary of the orthogonal convex hull, then the ball will instead alternate between both hooks (but will never leave the core).

The main idea of the rest of the construction is to limit the possible paths between the beacon and the core so that, in order to reach the left hook (via W_2), the beacon must walk counterclockwise relative to the core which makes the ball move to the right hook, and vice versa for the other hook. The shape of the pockets constrain the possible paths between the beacon and the core so that, even if we allow free movement in the exterior, the beacon cannot meet the ball.

2.2 Correctness

To be certain that there is no method for the two objects to meet (say, by going into a pocket then undoing the steps and going into another pocket), we verify all cases exhaustively as follows.

Define the ***grid space*** to be the two-dimensional cell complex consisting of ***cells***—corners, unit-length edges (excluding their endpoints), and unit-square faces (excluding their boundary)—of the unit square grid, and the ***beacon space*** and ***ball space*** to be the subspaces of the grid space that the beacon and ball can occupy, respectively. (The ball space consists only of vertices and edges, i.e., zero- and one-dimensional cells, of the grid space.) Because our example is orthogonal with boundaries on the unit square grid, as long as the beacon remains within one cell of its space, the ball will remain within one cell of its space (of lower or equal dimension), in particular because, when the ball rests on an edge of the grid, that edge is orthogonal to the line segment connecting the beacon and the ball. Define the ***configuration space*** to be the directed graph in which each node is a pair of a cell in the beacon space (***beacon location***) and a cell in the ball space (***ball location***), and we add a directed edge (u, v) if the beacon location of u is incident to the beacon location of v, and moving the beacon from the former to the latter causes the ball to move from the ball location of u to the (not necessarily incident) ball location of v. We want to explore this graph and determine whether there is a node (1) reachable from the starting position and (2) where the ball and beacon are in a common cell of the grid space.

Although there is a quadratic number of possible pairs of positions for the ball and beacon, only a very small amount is reachable from the starting position shown. We ***label*** each ball location with the two-dimensional cells incident to it. From the starting position, the ball can be incident to only fourteen such two-dimensional cells, denoted by letters A to N in Fig. 4. The transitions between these cells, where the ball location has multiple labels, correspond to points where the ball might jump locations, and thus must be considered carefully. For each reachable vertex of the configuration space, we give the same labels to the beacon location and color it accordingly in Fig. 4.

The four images that form Fig. 4 encompass the whole range of reachable configurations. The bottom-left image shows the possible locations of the ball in the left hook (labels A to F, each marked with a different color). For each such location, we have highlighted the possible beacon locations (using the same color). For example, if we know that the ball is in the left hook (labels A to F), and the beacon is in the topmost-rightmost corner, then it follows that the ball location must be solely labeled A. The other three images depict the remaining pairs of reachable positions for the ball and beacon, and the transitions ("portals") between these different parts of the configuration space. In each image, white locations are impossible for the beacon to reach if we enforce the ball position to have one of the labels in that image.

Although the beacon location does not uniquely determine the ball location, there are at most three possible locations for the ball for any given beacon

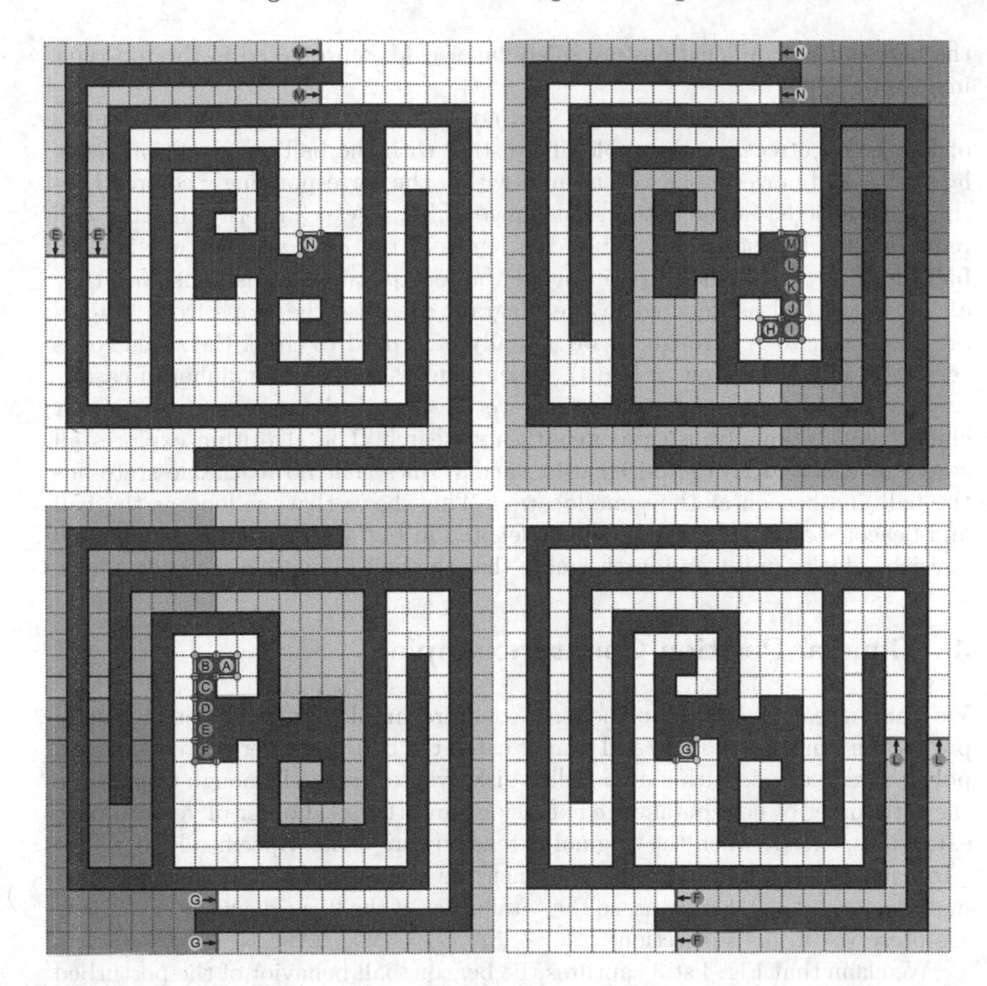

Fig. 4. Configuration space of pairs of positions reachable by the ball and beacon from the starting position in Fig. 3. Reachable ball locations are the colored vertices and edges in the core, where colors correspond to labels. The figure is split into four images, each with disjoint regions for the beacon (outlined by solid edges and colored differently). Circle-labeled arrows denote transitions ("portals") between regions from different images (along red edges, in the same direction as the arrow). The top and bottom rows are symmetric cases for the two hooks. These figures are computed automatically via a breadth-first search. (Color figure online)

location. There are also few movements of the beacon that cause a change in the ball location (cell): the beacon must transition between a colored zone and the boundary between colored zones (drawn with solid lines). For example, if we start in a configuration where the ball and beacon locations are solely labeled N, putting the beacon in the colored region in the top-left image of Fig. 4, and then the beacon moves sufficiently to the right, then we reach a configuration where the ball and beacon locations are solely labeled M, represented in the top-right image of Fig. 4.

Given the size of the instance, it is not difficult to verify that the images of Fig. 4 are correct. For example, if we start with the ball at a location solely labeled A and move the beacon around within the correspondingly colored beacon zone depicted in the bottom-left image of Fig. 4, we can verify that the ball remains in a location solely labeled A. Indeed, the only ways of moving the ball to a differently labeled position are the four pictured transitions, and they all move the ball as denoted in the diagram. Because the maps from H to N are symmetric to A through G, we actually only need to check the seven zones corresponding to one row of Fig. 4), which reduces to a small number of cases.

In addition to human verification, we have verified the correctness of our claim computationally using a breadth-first search. The algorithm explores all states reachable from the initial state, and verifies that no reachable state has the ball and beacon at the same location. This shows that, as long as the ball and beacon start in any configuration denoted in Fig. 4, the ball and beacon will never be able to unite, and furthermore that the ball cannot exit the two hooks.

3 General Position Counterexample

We claim we can modify our orthogonal counterexample into a polygon in general position in which the beacon still cannot catch the ball. Consider perturbing each polygon vertex to lie within an ε-radius disk, where ε is small enough to preserve the total order of horizontal edge extensions and the total order of vertical edge extensions, except when the original extensions are equal. Specifically, if we set ε to one thousandth of a grid square, then the angle of each edge changes by at most $\arctan \varepsilon < 0.001$ radians, so the total order of the 21 grid-square extensions are preserved in both directions.

We claim that Fig. 4 still captures the beacon/ball behavior of the perturbed example, with a perturbed notion of labeling. Specifically, we define the boundaries between ball labels B–F according to the extensions of the corresponding horizontal edges in the left hook, with the exception of the transition between labels D and E, which we define by the extension of the corresponding horizontal edge in the right hook (as there is no corresponding horizontal edge in the left hook). Symmetrically, we can define the boundaries between labels I–M. For vertical transitions, we define the boundaries between labels A and B and between labels F and G according to the extension of the corresponding vertical edge in the left hook (the right edge of cell C), and symmetrically between labels H and I and between labels M and N. The boundaries between the beacon colored zones are defined by the extensions of the same edges of the corresponding boundaries between ball labels. Since the perturbation is small and our example is designed so that each transition is affected by only one polygon edge (the one defined above), the resulting configuration space is the same graph as in the orthogonal polygon, with identical transition behavior. Therefore the same arguments hold.

Acknowledgments. We thank Dylan Hendrickson for helpful discussions, and the anonymous referees for helpful comments.

References

1. Biro, M.: Beacon-based routing and guarding. Ph.D. thesis, Stony Brook University, May 2013
2. Biro, M., Gao, J., Iwerks, J., Kostitsyna, I., Mitchell, J.S.: Beacon-based routing and coverage. In: Abstracts from the 21st Fall Workshop on Computational Geometry (2011)
3. Biro, M., Gao, J., Iwerks, J., Kostitsyna, I., Mitchell, J.S.: Combinatorics of beacon routing and coverage. In: Proceedings of the 25th Canadian Conference on Computational Geometry (CCCG 2013), Waterloo, Canada, August 2013
4. Biro, M., Iwerks, J., Kostitsyna, I., Mitchell, J.S.B.: Beacon-based algorithms for geometric routing. In: Dehne, F., Solis-Oba, R., Sack, J.-R. (eds.) WADS 2013. LNCS, vol. 8037, pp. 158–169. Springer, Heidelberg (2013). https://doi.org/10.1007/978-3-642-40104-6_14
5. Cleve, J., Mulzer, W.: Combinatorics of beacon-based routing in three dimensions. In: Bender, M.A., Farach-Colton, M., Mosteiro, M.A. (eds.) LATIN 2018. LNCS, vol. 10807, pp. 346–360. Springer, Cham (2018). https://doi.org/10.1007/978-3-319-77404-6_26
6. Kostitsyna, I., Kouhestani, B., Langerman, S., Rappaport, D.: An optimal algorithm to compute the inverse beacon attraction region. In: Speckmann, B., Tóth, C.D. (eds.) Proceedings of the 34th International Symposium on Computational Geometry (SoCG 2018), volume 99 of LIPIcs, pp. 55:1–55:14, Budapest, Hungary, June 2018
7. Kouhestani, B.: Efficient algorithms for beacon routing in polygons. Ph.D. thesis, Queen's University, Kingston, Canada (2017)
8. Kouhestani, B., Rappaport, D.: Edge patrolling beacon. In: Abstracts from the 20th Japan Conference on Discrete and Computational Geometry, Graphs, and Games, pp. 101–102 (2017)
9. Rappaport, D.: Personal communication held after the presentation of [8] (2017)
10. Shermer, T.C.: A combinatorial bound for beacon-based routing in orthogonal polygons. In: Proceedings of the 27th Canadian Conference on Computational Geometry (CCCG 2015), Kingston, Ontario, pp. 213–219 (2015)

Global Location-Domination in the Join and Cartesian Product of Graphs

Analen Malnegro(✉) and Gina Malacas

Iligan Institute of Technology, Mindanao State University,
Bonifacio Ave., Tibanga, 9200 Iligan City, Philippines
{analen.malnegro,gina.malacas}@g.msuiit.edu.ph

Abstract. In this paper, global locating-dominating sets of the join and Cartesian product of graphs are characterized. Global location-domination numbers and bounds of these graphs and some special graphs are determined. The relationship between the location-domination numbers of the join of two graphs and its complement is also shown.

Keywords: Domination · Location-domination · Global location-domination

1 Introduction

Let $G = (V(G), E(G))$ be a simple graph. The set $N_G(v) = \{u \in V(G) : uv \in E(G)\}$ is the *open neighborhood* of a vertex $v \in V(G)$. The set $N_G[v] = N_G(v) \cup \{v\}$ is the *closed neighborhood* of a vertex $v \in G$. If $X \subseteq V(G)$, then the set $N_G(X) = \bigcup_{v \in X} N_G(v)$ is the *open neighborhood* of X in G. The set $N_G[X] = N_G(X) \cup X$ is the *closed neighborhood* of X in G. The *complement* of a graph G, denoted by $\overline{G}$, is the graph with vertex set $V(\overline{G}) = V(G)$ and $E(\overline{G}) = \{uv : u, v \in V(G) \text{ and } uv \notin E(G)\}$. The *join* of two graphs G and H, denoted by $G + H$, is the graph with vertex set $V(G+H) = V(G) \cup V(H)$ and edge set $E(G+H) = E(G) \cup E(H) \cup \{uv : u \in V(G), v \in V(H)\}$. The *Cartesian product* of two graphs G and H, denoted by $G \Box H$, is the graph with vertex set $V(G \times H) = V(G) \times V(H)$ and edge set $E(G \times H)$ satisfying the following conditions: $(u_1, v_1)(u_2, v_2) \in E(G \times H)$ if and only if either $u_1u_2 \in E(G)$ and $v_1 = v_2$ or $u_1 = u_2$ and $v_1v_2 \in E(H)$.

Dominating sets and related topics have been studied extensively in the past few decades. A set $D \subseteq V(G)$ is a *dominating set* of G if for every $v \in V(G) \setminus D$, there exists $u \in D$ such that $uv \in E(G)$, that is, $N_G[D] = V(G)$. The *domination number* $\gamma(G)$ of G is the minimum cardinality of a dominating set of G. A dominating set is *global* if it is a dominating set of both G and its complement graph, $\overline{G}$. The minimum cardinality of a global dominating set of G is the *global domination number* of G, denoted by $\gamma_g(G)$.

Supported by DOST-ASTHRDP, Philippines.

J. Akiyama et al. (Eds.): JCDCGGG 2018, LNCS 13034, pp. 36–42, 2021.
https://doi.org/10.1007/978-3-030-90048-9_4

A subset S of $V(G)$ is a *locating set* in a connected graph G if for every two distinct vertices u and v of $V(G) \setminus S$, $N_G(u) \cap S \neq N_G(v) \cap S$.

There are problems on detection devices that can be modelled with graphs. Finding the minimum number of devices needed according to the type of devices and the necessity of locating the object gives rise to locating-dominating sets. A set $S \subseteq V(G)$ is a *locating-dominating set* (LD-set) if S is a dominating set such that for every two different vertices $u, v \in V(G) \setminus S, N_G(u) \cap S \neq N_G(v) \cap S$. The *location-domination number* $\lambda(G)$ of G is the minimum cardinality of a locating-dominating set. A set $S \subseteq V(G)$ is a *global locating-dominating set* if it is a locating-dominating set of both G and its complement, $\overline{G}$. The *global location-domination number* $\lambda_g(G)$ of G is the minimum cardinality of a global locating-dominating set of G.

The concepts of locating-dominating sets and related topics were discussed in [1,2,4]. An updated list of papers related to locating-dominating sets can be found in [6]. The concept of global domination was introduced by Sampathkumar [8]. Recently, the concept of global locating-dominating set is found in [3,5,7].

2 Preliminaries

This section presents some results needed in this paper.

Proposition 1. [8]

(i) For a graph G with p vertices, $\gamma_g(G) = p$ if and only if $G = K_p$ or $G = \overline{K}_p$.
(ii) $\gamma_g(K_{m,n}) = 2$ for all $m, n \geq 1$.
(iii) $\gamma_g(C_4) = 2$, $\gamma_g(C_5) = 3$ and $\gamma_g(C_n) = \lceil \frac{n}{3} \rceil$ for all $n \geq 6$.
(iv) $\gamma_g(P_n) = n$ for $n = 2, 3$ and $\gamma_g(P_n) = \lceil \frac{n}{3} \rceil$ for $n \geq 4$.

Lemma 1. [2] *Let G be a connected graph. If $\lambda(G) = 2$, then $3 \leq |V(G)| \leq 5$.*

Lemma 2. [5] *Let $G = (V, E)$ be a graph and $S \subseteq V$. If $x, y \in V \setminus S$, then $N_G(x) \cap S \neq N_G(y) \cap S$ if and only if $N_{\overline{G}}(x) \cap S \neq N_{\overline{G}}(y) \cap S$.*

Remark 1. [5] Every LD-set of a non-connected graph G is the union of LD-sets of its connected components and the location-domination number is the sum of the location-domination number of its connected components.

Proposition 2. [5] *For any graph G, $\lambda(G) \leq \lambda_g(G) \leq \lambda(G) + 1$.*

Lemma 3. [5] *If $n \geq 7$, then $\lambda(\overline{C_n}) = \lambda(\overline{P_n}) = \lambda(P_{n-1})$.*

Proposition 3. [5,9] *Let G be a graph of order $n \geq 1$. If G belongs to the set $\{P_n, C_n, W_n, K_n, K_{1,n-1}, K_{r,n-r}, K_2(r, n-r-2)\}$, then the values of $\lambda(G)$ and $\lambda(\overline{G})$ are known and they are displayed in Tables 1 and 2.*

Theorem 1. [5] *Let G be a graph of order $n \geq 1$. If G belongs to the set $\{P_n, C_n, W_n, K_n, K_{1,n-1}, K_{r,n-r}, K_2(r, n-r-2)\}$, then $\lambda_g(G)$ is known and it is displayed in Tables 1 and 2.*

Remark 2. For any graph $G = (V, E)$, $\gamma_g(G) \leq \lambda_g(G)$.

Table 1. The values of $\lambda(G)$, $\lambda(\overline{G})$ and $\lambda_g(G)$ of small paths, cycles and wheels.

G	P_1	P_2	P_3	P_4	P_5	P_6	C_4	C_5	C_6	W_5	W_6	W_7
$\lambda(G)$	1	1	2	2	2	3	2	2	3	2	3	3
$\lambda(\overline{G})$	1	2	2	2	2	3	2	2	3	3	3	4
$\lambda_g(G)=\lambda_g(\overline{G})$	1	2	2	2	3	3	2	3	3	3	3	4

Table 2. The values of $\lambda(G)$, $\lambda(\overline{G})$ and $\lambda_g(G)$ for some families of graphs.

G	P_n	C_n	W_n	K_n	$K_{1,n-1}$	$K_{r,n-r}$	$K_2(r,n-r-2)$
order n	$n\geq 7$	$n\geq 7$	$n\geq 8$	$n\geq 2$	$n\geq 4$	$2\leq r\leq n-r$	$2\leq r\leq n-r-2$
$\lambda(G)$	$\lceil\frac{2n}{5}\rceil$	$\lceil\frac{2n}{5}\rceil$	$\lceil\frac{2n-2}{5}\rceil$	$n-1$	$n-1$	$n-2$	$n-2$
$\lambda(\overline{G})$	$\lceil\frac{2n-2}{5}\rceil$	$\lceil\frac{2n-2}{5}\rceil$	$\lceil\frac{2n+1}{5}\rceil$	n	$n-1$	$n-2$	$n-3$
$\lambda_g(G)=\lambda_g(\overline{G})$	$\lceil\frac{2n}{5}\rceil$	$\lceil\frac{2n}{5}\rceil$	$\lceil\frac{2n+1}{5}\rceil$	n	$n-1$	$n-2$	$n-2$

3 Global Location-Domination of Some Graphs

This section presents the global location-domination number of some graphs.

Theorem 2. *Let $G=(V,E)$ be a connected graph. If $\lambda_g(G)=2$, then $2\leq |V(G)|\leq 4$.*

Proof. Suppose that $\lambda_g(G)=2$. Then $2\leq |V(G)|$. By Lemma 1, $|V(G)|\leq 5$. Hence, it is sufficient to show that $|V(G)|\neq 5$. Suppose that $|V(G)|=5$. Let $V(G)=\{v_1,v_2,v_3,v_4,v_5\}$. Let $S=\{v_2,v_4\}$ be a minimum global LD-set of G. Then S is a dominating set of both G and $\overline{G}$. Also, $N_G(v_1)\cap S=\{v_2\}$, $N_G(v_3)\cap S=\{v_2,v_4\}=S$, and $N_G(v_5)\cap S=\{v_4\}$. Since $N_G(v_3)\cap S=S$, it follows that $N_{\overline{G}}(v_3)\cap S=\emptyset$. This means that S is not a dominating set of $\overline{G}$, contrary to our assumption. Therefore, $|V(G)|\leq 4$. □

Theorem 3. *Let G be a graph of order $n\geq 2$. Then $\lambda_g(G)=n$ if and only if $G=K_n$ or $G=\overline{K}_n$.*

Proof. Suppose that $\lambda_g(G)=n$. Consider the following cases:

Case 1: Suppose G is connected. Suppose further that $G\neq K_n$. Then there exist $x,y\in V(G)$ such that $d_G(x,y)=2$, say $[x,z,y]$ is a path. Let $S=V(G)\setminus\{x\}$. Then, clearly, S is an LD-set of G. Now, since $xy\in E(\overline{G})$, S is also an LD-set of $\overline{G}$. Thus, $\lambda_g(G)\leq |S|=n-1$, contrary to our assumption. Therefore, $G=K_n$.

Case 2: Suppose G is disconnected. Let C be a component of G. Suppose that $|V(C)|\geq 2$. Then there exist $a,b\in V(C)$ such that $ab\in E(C)\subset E(G)$. Let C^* be a component of G with $C\neq C^*$. Then $av,bv\notin E(G)$ for all $v\in V(C^*)$. Thus, $S^*=V(G)\setminus\{a\}$ is a global LD-set of G, contrary to our assumption that $\lambda_g(G)=n$. Therefore, $|V(C)|=1$ for every component C of G, that is, $G=\overline{K_n}$.

Conversely, suppose that $G=K_n$ or $G=\overline{K}_n$. Then by Remark 2 and Proposition 1 (*i*), $\lambda_g(G)=n$. □

4 Join of Graphs

This section gives necessary and sufficient conditions for subsets of the join of graphs to be global locating-dominating. The corresponding global location-domination numbers of these graphs are also determined.

Theorem 4. *Let G be a graph and let $K_1 = \langle v \rangle$. A subset S of $V(K_1+G)$ is a global LD-set of K_1+G if and only if $S = \{v\} \cup S_{\overline{G}}$, where $S_{\overline{G}}$ is an LD-set of $\overline{G}$.*

Proof. Suppose that $S \subseteq V(K_1+G)$ is a global LD-set of K_1+G. Since v is an isolated vertex of $\overline{K_1+G} \cong K_1 \cup \overline{G}$ and S is a dominating set of $\overline{K_1+G}$, it follows that $v \in S$. Let $S_{\overline{G}} = V(\overline{G}) \cap S$. Since v does not dominate any vertex of $\overline{G}$ and S is a dominating set of $\overline{K_1+G}$, it follows that $S_{\overline{G}}$ is a dominating set of $\overline{G}$. Let $a, b \in V(\overline{G}) \setminus S_{\overline{G}}$ with $a \neq b$. Then

$$N_{\overline{G}}(a) \cap S_{\overline{G}} = N_{\overline{K_1+G}}(a) \cap S \neq N_{\overline{K_1+G}}(b) \cap S = N_{\overline{G}}(b) \cap S_{\overline{G}}.$$

Thus, $N_{\overline{G}}(a) \cap S_{\overline{G}} \neq N_{\overline{G}}(b) \cap S_{\overline{G}}$. Hence, $S_{\overline{G}}$ is a locating set of $\overline{G}$. Accordingly, S is an LD-set of $\overline{G}$.

Conversely, suppose that $S = \{v\} \cup S_{\overline{G}}$, where $S_{\overline{G}}$ is an LD-set of $\overline{G}$. Clearly, S is a dominating set of K_1+G and $\overline{K_1+G}$. Let $a, b \in V(\overline{K_1+G}) \setminus S$. Then $a, b \in V(\overline{G}) \setminus S_{\overline{G}}$ and $N_{\overline{G}}(a) \cap S_{\overline{G}} \neq N_{\overline{G}}(b) \cap S_{\overline{G}}$. Hence,

$$N_{\overline{K_1+G}}(a) \cap S = N_{\overline{G}}(a) \cap S_{\overline{G}} \neq N_{\overline{G}}(b) \cap S_{\overline{G}} = N_{\overline{K_1+G}}(b) \cap S.$$

This shows that S is an LD-set of $\overline{K_1+G}$. By Lemma 2, $N_{K_1+G}(a) \cap S \neq N_{K_1+G}(b) \cap S$. Thus, S is an LD-set of K_1+G. Therefore, S is a global LD-set of K_1+G. □

Corollary 1. *Let G be a graph and let $K_1 = \langle v \rangle$. A subset S of $V(K_1+G)$ is a minimum global LD-set of K_1+G if and only if $S = \{v\} \cup S_{\overline{G}}$, where $S_{\overline{G}}$ is a minimum LD-set of $\overline{G}$. In particular, $\lambda_g(K_1+G) = 1 + \lambda(\overline{G})$.*

Theorem 5. *Let G and H be non-trivial graphs. Then $S \subseteq V(G+H)$ is a global LD-set of $G+H$ if and only if $S = S_G \cup S_H$, where S_G and S_H are LD-sets of $\overline{G}$ and $\overline{H}$, respectively.*

Proof. Suppose that $S \subseteq V(G+H)$ is a global LD-set of $G+H$. Let $S_G = V(\overline{G}) \cap S$ and $S_H = V(\overline{H}) \cap S$. Since $S_G \subseteq V(\overline{G})$ and S is a dominating set of $\overline{G+H} \cong \overline{G} \cup \overline{H}$, it follows that S_G is a dominating set of $\overline{G}$. Let $x, y \in V(\overline{G})$ with $x \neq y$. Since S is an LD-set of $\overline{G+H}$,

$$N_{\overline{G}}(x) \cap S_G = N_{\overline{G+H}}(x) \cap S \neq N_{\overline{G+H}}(y) \cap S = N_{\overline{G}}(y) \cap S_G.$$

Thus, S_G is an LD-set of $\overline{G}$. Similarly, S_H is an LD-set of $\overline{H}$.

Conversely, suppose $S = S_G \cup S_H$, where S_G and S_H are LD-sets of $\overline{G}$ and $\overline{H}$, respectively. Let $a, b \in V(\overline{G+H}) \setminus S$ with $a \neq b$. If $a, b \in V(\overline{G})$, then

$N_{\overline{G+H}}(a) \cap S = N_{\overline{G}}(a) \cap S_G \neq N_{\overline{G}}(b) \cap S_G = N_{\overline{G+H}}(b) \cap S$. Similarly, $N_{\overline{G+H}}(a) \cap S \neq N_{\overline{G+H}}(b) \cap S$ if $a, b \in V(\overline{H})$. Suppose $a \in V(\overline{G})$ and $b \in V(\overline{H})$. Then $N_{\overline{G+H}}(a) \cap S = N_{\overline{G}}(a) \cap S_G \neq N_{\overline{H}}(b) \cap S_H = N_{\overline{G+H}}(b) \cap S$ since $\overline{G+H} \cong \overline{G} \cup \overline{H}$. Accordingly, S is a locating set of $\overline{G+H}$. Since $N_{\overline{G+H}}(a) \cap S \neq N_{\overline{G+H}}(b) \cap S$ implies that $N_{G+H}(a) \cap S \neq N_{G+H}(b) \cap S$ by Lemma 2, S is a locating set of $G+H$. Clearly, S is a dominating set of $G+H$ and $\overline{G+H}$. Therefore, S is a global LD-set of $G+H$. □

The following result is immediate from Theorem 5.

Corollary 2. *Let G and H be non-trivial graphs. Then $S \subseteq V(G+H)$ is a minimum global LD-set of $G+H$ if and only if $S = S_G \cup S_H$, where S_G and S_H are minimum LD-sets of $\overline{G}$ and $\overline{H}$, respectively.*

Theorem 6. *Let G and H be two graphs. Then $\lambda(\overline{G+H}) = \lambda(\overline{G}) + \lambda(\overline{H})$.*

Proof. Note that $\overline{G+H} \cong \overline{G} \cup \overline{H}$. Since $\overline{G} \cup \overline{H}$ is a non-connected graph, by Remark 1, $\lambda(\overline{G+H}) = \lambda(\overline{G} \cup \overline{H}) = \lambda(\overline{G}) + \lambda(\overline{H})$. □

Theorem 7. *Let G and H be two graphs. Then $\lambda_g(G+H) = \lambda(\overline{G}) + \lambda(\overline{H})$.*

Proof. Let S be a minimum global LD-set of $G+H$. Then by Corollary 2, $S = S_G \cup S_H$, where S_G and S_H are minimum LD-sets of $\overline{G}$ and $\overline{H}$, respectively. Since S_G and S_H are disjoint sets,

$$\lambda_g(G+H) = |S| = |S_G| + |S_H| = \lambda(\overline{G}) + \lambda(\overline{H}).$$

□

Corollary 3. *Let G and H be any graphs. Then $\lambda(G+H) \leq \lambda(\overline{G}) + \lambda(\overline{H})$.*

Proof. By Proposition 2 and Theorem 7,

$$\lambda(G+H) \leq \lambda_g(G+H) = \lambda(\overline{G}) + \lambda(\overline{H}).$$

□

From Theorems 5–7 and Theorem 1, the next results are immediate.

Corollary 4. *Let G be a graph and let K_n be the complete graph with $n \geq 1$. Then $\lambda_g(K_n + G) = n + \lambda(\overline{G})$.*

Corollary 5. *Let $F_{m,n} = \overline{K}_m + P_n$, where m and n are positive integers, and $m \geq 2$. Then*

$$\lambda_g(F_{m,n}) = \begin{cases} (m-1)+2 & 2 \leq n \leq 5 \\ (m-1)+3 & n = 6 \\ (m-1)+\lceil \frac{2n-2}{5} \rceil & n \geq 7. \end{cases}$$

Corollary 6. *Let $W_{m,n} = \overline{K}_m + C_n$, where m and n are positive integers, and $m \geq 2$. Then*

$$\lambda_g(W_{m,n}) = \begin{cases} (m-1)+2 & 4 \leq n \leq 5 \\ (m-1)+3 & n = 6 \\ (m-1)+\lceil \frac{2n-2}{5} \rceil & n \geq 7. \end{cases}$$

5 Cartesian Product of Graphs

This section presents a characterization of a particular type of global locating-dominating set of the Cartesian product of two graphs. An upper bound on the global location-domination number of this graph is also given.

Theorem 8. *Let G and H be non-trivial connected graphs. Then $S = A \times B$, where $A \subseteq V(G)$ and $B \subseteq V(H)$, is a global LD-set of $G\Box H$ if and only if either*

(i) $A = V(G)$ and B is an LD-set of H; or
(ii) $B = V(H)$ and A is an LD-set of G.

Proof. Suppose that $A = V(G)$. Let $a \in V(H) \setminus B$ and let $u \in A$. Then $(u, a) \notin A \times B$. Hence, there exists $(v, b) \in A \times B$ such that $(u, a)(v, b) \in E(G\Box H)$. Since $(v, a) \notin A \times B$, $u = v$ and $ab \in E(H)$. Thus, B is a dominating set of H. If B is not a locating set of H, then there exist $x, y \in V(H) \setminus B$ such that $N_H(x) \cap B = N_H(y) \cap B$. Let $T = N_H(x) \cap B$. Then $(\{u\} \times T) \subseteq S$. Since $N_{G\Box H}((u, x)) \cap S = (\{u\} \times T) = N_{G\Box H}((u, y)) \cap S$, it follows that S is not an LD-set of $G\Box H$. This is also a contradiction to our assumption regarding S. Thus, B is a locating set of H. Accordingly, B is an LD-set of H.

Next, suppose that $A \neq V(G)$. Suppose further that $B \neq V(H)$. Pick $p \in V(G) \setminus A$ and $q \in V(H) \setminus B$. Then $(p, t) \notin A \times B$ and $(v, q) \notin A \times B$ for all $t \in B$ and $v \in A$. This implies that (p, q) is not dominated by any element of $A \times B$, a contradiction. Thus, $B = V(H)$. Hence, A must be an LD-set of G.

Conversely, suppose that $A = V(G)$ and B is an LD-set of H or $B = V(H)$ and A is an LD-set of G. Assume, without loss of generality, that $A = V(G)$ and B is an LD-set of H. Suppose that S is not a global LD-set of $G\Box H$. Then S is not an LD-set of $G\Box H$ or $\overline{G\Box H}$. Suppose S is not an LD-set of $G\Box H$. Then S is not a dominating set of $G\Box H$ or S is not a locating set of $G\Box H$. If S is not a dominating set of $G\Box H$, then there exists $(u, x) \in V(G\Box H) \setminus S$ such that $N_{G\Box H}((u, x)) \cap S = \emptyset$. Since $\{u\} \times (N_H(x) \cap B) = N_{G\Box H}((u, x)) \cap S = \emptyset$, $N_H(x) \cap B = \emptyset$. That is, $A \neq V(G)$ or B is not a dominating set of H. This contradicts our assumption. Thus, S is a dominating set of $G\Box H$. If S is not a locating set of $G\Box H$, then there exist $(u, a), (v, b) \in V(G\Box H) \setminus S$ such that $N_{G\Box H}((u, a)) \cap S = N_{G\Box H}((v, b)) \cap S$. Since

$$\{u\} \times (N_H(a) \cap B) = N_{G\Box H}((u, a)) \cap S = N_{G\Box H}((v, b)) \cap S = \{v\} \times (N_H(b) \cap B),$$

it follows that $u = v$ and $N_H(a) \cap B = N_H(b) \cap B$. This contradicts the fact that B is a locating set of H. Hence, S is a locating set of $G\Box H$. Accordingly, S is an LD-set of $G\Box H$. Similarly, S is an LD-set of $\overline{G\Box H}$. Thus, S is a global LD-set of $G\Box H$. □

The next result is immediate from Theorem 8.

Corollary 7. *Let G and H be non-trivial connected graphs. Then*

$$\lambda_g(G\Box H) \leq \min\{|V(G)|\lambda(H), |V(H)|\lambda(G)\}.$$

References

1. Cáceres, J., Hernando, C., Mora, M., Pelayo, I.M., Puertas, M.L.: Locating-dominating codes: bounds and extremal cardinalities. Appl. Math. Comput. **220**, 38–45 (2013)
2. Canoy, S., Jr., Malacas, G.: Determining the intruder's location in a given network: locating-dominating sets in a graph. NRCP Res. J. **13**(1), 1–8 (2013)
3. Hernando, C., Mora, M., Pelayo, I.M.: Locating domination in bipartite graphs and their complements. Discret. Appl. Math. **263**, 195–203 (2019)
4. Hernando, C., Mora, M., Pelayo, I.M.: Nordhaus-Gaddum bounds for locating domination. Eur. J. Comb. **36**, 1–6 (2014)
5. Hernando, C., Mora, M., Pelayo, I.M.: On global location-domination in graphs. Ars Mathematica Contemporanea **8**, 365–379 (2015)
6. Lobstein, A.: Watching systems, identifying, locating-dominating and discriminating codes in graphs. https://www.lri.enst.fr/lobstein/debutBIBidetlocdom.pdf
7. Malacas, G., Malnegro, A.: Global location-domination in the lexicographic product and corona of graphs. Adv. Appl. Discret. Math. **20**(1), 61–71 (2019). https://doi.org/10.17654/DM020010061
8. Sampathkumar, E.: The global domination number of a graph. J. Math. Phys. Sci. **23**, 377–385 (1989)
9. Slater, P.J.: Dominating and reference sets in a graph. J. Math. Phys. Sci. **22**, 445–455 (1988)

The Metric Dimension of the Join of Paths and Cycles

Ian June L. Garces[1] and Jose B. Rosario[2(✉)]

[1] Ateneo de Manila University, Quezon City, Philippines
ijlgarces@ateneo.edu
[2] Isabela State University, Cabagan Campus, Cabagan, Isabela, Philippines
jose.b.rosario@isu.edu.ph

Abstract. For an ordered subset $W = \{w_1, w_2, \ldots, w_k\}$ of vertices in a connected graph G and a vertex v of G, the metric representation of v with respect to W is the k-vector $r(v|W) = (d(v, w_1), \ldots, d(v, w_k))$, where $d(v, w_i)$ is the distance of the vertices v and w_i in G. The set W is called a resolving set of G if $r(u|W) = r(v|W)$ implies $u = v$. A resolving set of G with minimum cardinality is called a metric basis of G. The metric dimension of G, denoted by $\beta(G)$, is the cardinality of a metric basis of G.

The *join* of G and H, denoted by $G + H$, is the graph with vertex set $V(G+H) = V(G) \cup V(H)$ and edge set $E(G+H) = E(G) \cup E(H) \cup \{uv : u \in V(G), v \in V(H)\}$. In general, the join of m graphs $G_1, G_2, \ldots, G_m$, denoted by $G = G_1 + G_2 + \cdots + G_m$, has vertex set $V(G) = V(G_1) \cup V(G_2) \cup \cdots \cup V(G_m)$ and edge set $E(G) = E(G_1) \cup E(G_2) \cup \cdots \cup E(G_m) \cup \{uv : u \in G_i, v \in G_j, i \neq j\}$.

In this paper, we compute the metric dimension of the join of a finite number of paths, the join of a finite number of cycles, and the join of paths and cycles.

Keywords: Join of graphs · Metric basis · Metric dimension · Resolving set

1 Introduction

Let $G = (V(G), E(G))$ be a finite, simple, and connected graph. The *distance* between two vertices u and v of G, denoted by $d(u, v)$, is the length of the shortest $u - v$ path in G. For an ordered subset $W = \{w_1, w_2, \ldots, w_k\}$ of $V(G)$, we refer to the k-vector (ordered k-tuple) $r(v|W) = (d(v, w_1), \ldots, d(v, w_k))$ as the *metric representation of vertex v with respect to W*. The set W is called a *resolving set of G* if every two distinct vertices u and v satisfy $r(u|W) \neq r(v|W)$. A *metric basis* of G is a resolving set of G with minimum cardinality, and the *metric dimension of G* refers to this cardinality, denoted by $\beta(G)$.

For a given ordered set $W = \{w_1, w_2, \ldots, w_k\}$ of vertices of G, it is not difficult to see that the ith component of $r(v|W)$ is 0 if and only if $v = w_i$. Thus, to show that W is a resolving set, it suffices to verify that $r(u|W) \neq r(v|W)$ for each pair of distinct vertices $u, v \in V(G) \backslash W$.

J. Akiyama et al. (Eds.): JCDCGGG 2018, LNCS 13034, pp. 43–58, 2021.
https://doi.org/10.1007/978-3-030-90048-9_5

Remark 1. Although W is treated as an ordered set, when its elements are permuted, the coordinates of $r(v|W)$ will follow correspondingly.

The metric dimension problem was first introduced by Harary and Melter [4] in 1976 and independently by Slater [6] in 1975. Several authors studied this topic and published numerous results. Chartrand et al. [3] proved the following characterizations of connected graphs of order $n \geq 2$ with metric dimension 1, $n-1$ and $n-2$.

Theorem 1 (Chartrand et al. [3]). *Let G be a connected graph of order $n \geq 2$. Then*

(i) $\beta(G) = 1$ *if and only if* $G = P_n$.
(ii) $\beta(G) = n-1$ *if and only if* $G = K_n$.
(iii) *for* $n \geq 4$, $\beta(G) = n-2$ *if and only if* $G = K_{s,t}$ $(s,t \geq 1)$, $G = K_s + \overline{K_t}$ $(s \geq 1, t \geq 2)$, *or* $G = K_s + (K_1 \cup K_t)$ $(s,t \geq 1)$.

The *join* of two graphs G and H, denoted by $G+H$, is the graph with vertex set $V(G+H) = V(G) \cup V(H)$ and edge set $E(G+H) = E(G) \cup E(H) \cup \{uv : u \in V(G), v \in V(H)\}$. In general, the join of m graphs $G_1, G_2, \ldots, G_m$, denoted by $G = G_1 + G_2 + \cdots + G_m$, has vertex set $V(G) = V(G_1) \cup V(G_2) \cup \cdots \cup V(G_m)$ and edge set $E(G) = E(G_1) \cup E(G_2) \cup \cdots \cup E(G_m) \cup \{uv : u \in V(G_i), v \in V(G_j), i \neq j\}$.

In [5], Shanmukha et al. developed a formula for the metric dimension of the join of the complete graph K_1 with a cycle C_n, $n \geq 3$, known to be the wheel graph on $n+1$ vertices which we denote here by $W_{1,n}$ while Cáceres et al. [2] determined the metric dimension of the join of the complete graph K_1 with a path P_n, $n \geq 2$, known to be the fan graph on $n+1$ vertices, which we denote here by $F_{1,n}$. These results are as follows:

Theorem 2 (Shanmukha et al. [5]). (i) $\beta(W_{1,3}) = \beta(W_{1,6}) = 3$.
(ii) $\beta(W_{1,4}) = \beta(W_{1,5}) = 2$.
(iii) $\beta(W_{1,x+5k}) = \begin{cases} 3+2k, \text{ where } x = 7 \text{ or } 8, \text{ all } k = 0,1,2,\ldots \\ 4+2k, \text{ where } x = 9 \text{ or } 10 \text{ or } 11, \text{ all } k = 0,1,2,\ldots \end{cases}$

For our convenience and for computational simplicity, we adapt the formula $\beta(W_{1,n}) = \beta(K_1 + C_n) = \lfloor \frac{2n+2}{5} \rfloor$ obtained by Buczkowski et al. in [1] for $n \geq 7$.

Theorem 3 (Cáceres et al. [2]). *Let $n \geq 2$ be an integer. Then*

$$\beta(F_{1,n}) = \beta(K_1 + P_n) = \begin{cases} 2, & \text{if } n = 2,3 \\ 3, & \text{if } n = 6 \\ \lfloor \frac{2n+2}{5} \rfloor, & \text{otherwise.} \end{cases}$$

Remark 2. As mentioned in [5], if v is the central vertex of the wheel graph $W_{1,n} = K_1 + C_n$, then v does not belong to any metric basis. This implies that, if S is a resolving set of $W_{1,n} = K_1 + C_n$, then $S \subset V(C_n)$. This also holds true for the fan graph $F_{1,n} = K_1 + P_n$ for $n \geq 7$, that is, if S is a resolving set of $F_{1,n} = K_1 + P_n$, then $S \subset V(P_n)$.

We state following theorem by Sooryanarayana (Theorem 7 in [7]) which is our basis in the formulation of Remark 4 in the next section.

Theorem 4 (Sooryanarayana [7]). *A graph G with $\beta(G) = k$, cannot contain a subgraph isomorphic to $K_{2^k+1} - (2^{k-1} - 1)e$ as a subgraph of G.*

In this paper, we compute the metric dimension of the join of a finite number of paths, the join of a finite number of cycles, and the join of paths and cycles.

2 The Join of a Finite Number of Paths

For integers $m, n \geq 2$, let $P_{m,n} = P_m + P_n$, where P_m and P_n are paths with vertices $u_1, u_2, \ldots, u_m$ and $v_1, v_2, \ldots, v_n$, respectively. We note that the fan graphs $F_{1,m} \cong v_j + P_m$ for each $v_j \in V(P_n)$ and $F_{1,n} \cong u_i + P_n$ for each $u_i \in V(P_m)$ are subgraphs of $P_{m,n}$. We also note that, by symmetry, $P_m + P_n = P_n + P_m$. Thus, to compute the metric dimension of $P_{m,n}$, it is sufficient to consider the cases $m = n$ and $m < n$.

Remark 3. Let G_1 and G_2 be two finite, simple, undirected, and connected graphs with $|V(G_1)|, |V(G_2)| \geq 2$. If S is a resolving set for $G_1 + G_2$, then $S = S_{G_1} \cup S_{G_2}$, where S_{G_1} and S_{G_2} are nonempty subsets of $V(G_1)$ and $V(G_2)$, respectively. In general, if S is a resolving set for $G = G_1 + G_2 + \cdots + G_m$, where G_i is a finite, simple, undirected, and connected graph and $|V(G_i)| \geq 2$ for each $1 \leq i \leq m$, $m \geq 2$, then $S = \cup_{i=1}^{m} S_{G_i}$, where S_{G_i} is a nonempty subset of $V(G_i)$.

Remark 4. When $m = n = 2$, the graph $P_m + P_n \cong K_4$ and hence by Theorem 1, $\beta(P_m + P_n) = 3$. When $n \geq 3$, the graph $P_m + P_n$ contains a graph $P_2 + P_3$ which is isomorphic to $K_5 - e$ and hence by Theorem 4, $\beta(P_m + P_n) \geq 3$.

Proposition 1. *If $m, n \in \{2, 3, 4, 5, 6\}$, then*

$$\beta(P_{m,n}) = \begin{cases} 3, \text{ if } 2 \leq m \leq 3 \text{ and } 2 \leq n \leq 5 \\ 4, \text{ if } 2 \leq m \leq 3 \text{ and } n = 6, \text{ or } 4 \leq m \leq 5 \text{ and } 4 \leq n \leq 6 \\ 5, \text{ if } m = n = 6. \end{cases}$$

Proof. We consider four cases.

Case 1. $2 \leq m \leq 3$ and $2 \leq n \leq 5$

Note that $P_2 + P_2 \cong K_4$. Therefore, $\beta(P_2 + P_2) = 3$ by Theorem 1. One can easily check that the set $S = \{u_1, v_1, v_2\}$ is a resolving set for $P_m + P_n$ ($m = 2, 3$ and $n = 3, 4$) and the set $S = \{u_1, v_1, v_5\}$ is a resolving set for $P_m + P_n$ ($m = 2, 3$ and $n = 5$). Thus, we have $\beta(P_m + P_n) \leq 3$. Therefore, by Remark 4, we have $\beta(P_m + P_n) = 3$.

Case 2. $2 \leq m \leq 3$ and $n = 6$

It is not difficult to verify that the set $S = \{u_1, v_1, v_5, v_6\}$ is a resolving set for both $P_2 + P_6$ and $P_3 + P_6$. It follows that $\beta(P_2 + P_6) = \beta(P_3 + P_6) \leq 4$. To prove the reverse inequality, it suffices to prove that there exists no resolving set for $P_2 + P_6$ and $P_3 + P_6$ of cardinality 3 by Remark 4.

Suppose S' is a resolving set for P_2+P_6 and P_3+P_6 of cardinality 3. Then, by Remark 3, $S' = \{u_i\} \cup \{v_j, v_k\}$ for some $u_i \in V(P_m)$ and for some $v_j, v_k \in V(P_6)$ or $S' = \{u_i, u_l\} \cup \{v_j\}$ for some $u_i, u_l \in V(P_m)$ and for some $v_j \in V(P_6)$.

Suppose $S' = \{u_i\} \cup \{v_j, v_k\}$ for some $u_i \in V(P_m)$ and for some $v_j, v_k \in V(P_6)$ with $j < k$. If $k = j + 1$, then either $r(v_5|S') = r(v_6|S')$ or $r(v_1|S') = r(v_6|S')$ or $r(v_1|S') = r(v_2|S')$. If $k = j+2$, then either $r(u_{i+1}|S') = r(v_{j+1}|S')$, where $i + 1$ is computed modulo 2 or modulo 3. If $k = j + 3$, then either $r(v_3|S') = r(v_5|S')$ or $r(v_4|S') = r(v_6|S')$ or $r(v_2|S') = r(v_4|S')$. If $k = j + 4$, then either $r(v_4|S') = r(v_6|S')$ or $r(v_1|S') = r(v_3|S')$. If $k = j + 5$, then $r(v_3|S') = r(v_4|S')$. Now, suppose $S' = \{u_i, u_l\} \cup \{v_j\}$ for some $u_i, u_l \in V(P_m)$ and for some $v_j \in V(P_6)$. If $j = 1$ or $j = 6$, then $r(v_3|S') = r(v_4|S')$. If $2 \leq j \leq 5$, then $r(v_{j-1}|S') = r(v_{j+1}|S')$. In either case, there exist two vertices in $V(P_6)\backslash\{v_j, v_k\}$ or $V(P_6)\backslash\{v_j\}$ with the same metric representation with respect to S', contradicting the assumption that S' is a resolving set. Thus, $\beta(P_2+P_6) \geq 4$ and $\beta(P_3+P_6) \geq 4$. Therefore, we conclude that $\beta(P_2+P_6) = \beta(P_3+P_6) = 4$.

CASE 3. $4 \leq m \leq 5$ and $4 \leq n \leq 6$

It is not difficult to verify that the set $S = \{u_1, u_m, v_2, v_4\}$ is a resolving set for $P_{m,n}$. It follows that $\beta(P_{m,n}) \leq 4$. To prove the reverse inequality, it suffices to prove that there exists no resolving set for $P_{m,n}$ of cardinality 3 by Remark 4.

Suppose S' is a resolving set for $P_{m,n}$ of cardinality 3. Then, by Remark 3, $S' = \{u_i\} \cup \{v_j, v_k\}$ for some $u_i \in V(P_m)$ and for some $v_j, v_k \in V(P_n)$ or $S' = \{u_i, u_l\} \cup \{v_j\}$ for some $u_i, u_l \in V(P_m)$ and for some $v_j \in V(P_n)$.

Suppose $S' = \{u_i\} \cup \{v_j, v_k\}$ for some $u_i \in V(P_m)$ and for some $v_j, v_k \in V(P_n)$. If $i = 1$, then $r(u_3|S') = r(u_4|S')$. If $2 \leq i \leq m - 1$, then $r(u_{i-1}|S') = r(u_{i+1}|S')$. If $i = m$, then $r(u_1|S') = r(u_2|S')$. Now, suppose $S' = \{u_i, u_l\} \cup \{v_j\}$ for some $u_i, u_l \in V(P_m)$ and for some $v_j \in V(P_n)$. If $j = 1$, then $r(v_3|S') = r(v_4|S')$. If $2 \leq j \leq n-1$, then $r(v_{j-1}|S') = r(v_{j+1}|S')$. If $j = n$, then $r(v_1|S') = r(v_2|S')$. In any case, there exist two vertices in $V(P_m)\backslash\{u_i\}$ or $V(P_n)\backslash\{v_j\}$ with the same metric representation with respect to S', contradicting the assumption that S' is a resolving set. Thus, $\beta(P_{m,n}) \geq 4$. Therefore, we conclude that $\beta(P_{m,n}) = 4$.

CASE 4. $m = n = 6$

It is not difficult to verify that the set $S = \{u_2, u_4, v_2, v_4, v_6\}$ is a resolving set for $P_{6,6}$. It follows that $\beta(P_{6,6}) \leq 5$. To prove the reverse inequality, we show that there exists no resolving set for $P_{6,6}$ of cardinality 3 or 4 by Remark 4.

Suppose S' is a resolving set for $P_{m,n}$ of cardinality 3. Then, by Remark 3, $S' = \{u_i\} \cup \{v_j, v_k\}$ for some $u_i \in V(P_m)$ and for some $v_j, v_k \in V(P_n)$ or $S' = \{u_i, u_l\} \cup \{v_j\}$ for some $u_i, u_l \in V(P_m)$ and for some $v_j \in V(P_n)$.

Suppose $S' = \{u_i\} \cup \{v_j, v_k\}$ for some $u_i \in V(P_m)$ and for some $v_j, v_k \in V(P_n)$. If $i = 1$, then $r(u_3|S') = r(u_4|S')$. If $2 \leq i \leq 5$, then $r(u_{i-1}|S') = r(u_{i+1}|S')$. If $i = 6$, then $r(u_1|S') = r(u_2|S')$. Now, suppose $S' = \{u_i, u_l\} \cup \{v_j\}$ for some $u_i, u_l \in V(P_m)$ and for some $v_j \in V(P_n)$. If $j = 1$, then $r(v_3|S') = r(v_4|S')$. If $2 \leq j \leq 5$, then $r(v_{j-1}|S') = r(v_{j+1}|S')$. If $j = 6$, then $r(v_1|S') = r(v_2|S')$. In any case, there exist two vertices in $V(P_m)\backslash\{u_i\}$ or $V(P_n)\backslash\{v_j\}$ with the same metric representation with respect to

S', contradicting the assumption that S' is a resolving set. Thus, there exists no resolving set for $P_{6,6}$ of cardinality 3.

Now, suppose S' is a resolving set for $P_{m,n}$ of cardinality 4. It is sufficient to consider the case when $S' = \{u_i, u_j\} \cup \{v_k, v_l\}$ for some $u_i, u_j \in V(P_m)$ and for some $v_k, v_l \in V(P_n)$. Following the same arguments as in Case 2 and Case 3, it is not difficult to check that there exist two vertices in $V(P_m)\backslash\{u_i, u_j\}$ or $V(P_n)\backslash\{v_k, v_l\}$ with the same metric representation with respect to S' or there exists a vertex in $V(P_m)\backslash\{u_i, u_j\}$ and a vertex in $V(P_n)\backslash\{v_k, v_l\}$ with the same metric representation with respect to S'. In any case, we arrive at a contradiction. Thus, there exists no resolving set for $P_{6,6}$ of cardinality 4. Therefore, $\beta(P_{6,6}) \geq 5$ and we conclude that $\beta(P_{6,6}) = 5$. □

Proposition 2. *Let $m \in \{2,3,4,5,6\}$ and $n \geq 7$ be an integer. Then*

$$\beta(P_{m,n}) = \begin{cases} 1 + \beta(F_{1,n}), \text{ if } m = 2,3 \\ 2 + \beta(F_{1,n}), \text{ if } m = 4,5,6. \end{cases}$$

Proof. As noted above, $F_{1,n} \cong u_i + P_n \subset P_{m,n}$ for each vertex $u_i \in V(P_m)$. By Theorem 3, we have $\beta(F_{1,n}) = \beta(u_i + P_n) = \lfloor \frac{2n+2}{5} \rfloor$ for some vertex $u_i \in V(P_m)$. Let $S_{1,n} = \{w_1, w_2, \ldots, w_\alpha\}$ be a metric basis for $u_i + P_n$ where $\alpha = \lfloor \frac{2n+2}{5} \rfloor$. By Remark 2, $S_{1,n}$ is a subset of $V(P_n)$. It follows that $r(v_i|S_{1,n}) \neq r(v_j|S_{1,n})$ for every pair of distinct vertices $v_i, v_j \in V(P_n)\backslash S_{1,n}$. Note that $r(v_j|S_{1,n})$ consists of σ 1's and $(\alpha - \sigma)$ 2's where $0 \leq \sigma \leq 2$ for each $v_j \in V(P_n)\backslash S_{1,n}$.

Consider first the case when $m = 2$. Let $S = \{u_1\} \cup S_{1,n}$. By Remark 1, we can let $S = \{u_1, w_1, \ldots, w_\alpha\}$. Then we have $r(u_2|S) = (1, \underbrace{1, \ldots, 1}_{\alpha \text{ times}})$. Observe that for each distinct vertices $v_i, v_j \in V(P_n)\backslash S_{1,n}$, we have $r(v_i|S) = (1, d(v_i, w_1), \ldots, d(v_i, w_\alpha))$ and $r(v_j|S) = (1, d(v_j, w_1), \ldots, d(v_j, w_\alpha))$.

Since $r(v_i|S_{1,n}) \neq r(v_j|S_{1,n})$, it follows that $r(v_i|S) \neq r(v_j|S)$. Also, since $r(v_j|S_{1,n})$ consists of σ 1's and $(\alpha - \sigma)$ 2's where $0 \leq \sigma \leq 2$ for each $v_j \in V(P_n)\backslash S_{1,n}$, we have $r(u_2|S) \neq r(v_j|S)$ for each $v_j \in V(P_n)\backslash S_{1,n}$. Hence, the metric representations of the vertices of $V(P_{2,n})\backslash S$ with respect to S are distinct. Therefore, S is a resolving set and we conclude that $\beta(P_{2,n}) \leq 1 + \alpha$.

To prove that $\beta(P_{2,n}) \geq 1+\alpha$, it suffices to show that there exists no resolving set of cardinality less than or equal to $\alpha = \beta(F_{1,n})$.

Suppose S' is a resolving set for $P_{2,n}$ of cardinality less than or equal to α. By Remark 3, $S' = S_{P_2} \cup S_{P_n}$ where S_{P_2} and S_{P_n} are nonempty subsets of $V(P_2)$ and $V(P_n)$, respectively. Since $|S_{P_2}| \geq 1$, then $|S_{P_n}| \leq \alpha - 1$. By Theorem 3, there exist $v_i, v_j \in V(P_n)\backslash S_{P_n}$ such that $r(v_i|S_{P_n}) = r(v_j|S_{P_n})$. Consequently, $r(v_i|S') = r(v_j|S')$, which is a contradiction. Hence, there exists no resolving set for $P_{2,n}$ of cardinality less than or equal to α. Therefore, $\beta(P_{2,n}) \geq 1 + \alpha$ and we conclude that $\beta(P_{2,n}) = 1 + \beta(F_{1,n})$.

Following the same arguments above, we see that the set $S = \{u_1\} \cup S_{1,n}$ is also a metric basis for $P_{3,n}$. Hence, we conclude that $\beta(P_{3,n}) = 1 + \beta(F_{1,n})$. Furthermore, the set $S = \{u_2, u_4\} \cup S_{1,n}$ is a metric basis for $P_{m,n}$, $4 \leq m \leq 6$ and $n \geq 7$. Hence, we conclude that $\beta(P_{m,n}) = 2 + \beta(F_{1,n})$. □

Theorem 5. *Let $m, n \geq 7$ be integers. Then $\beta(P_{m,n}) = \beta(F_{1,m}) + \beta(F_{1,n})$.*

Proof. Again, we note that the fan graphs $F_{1,m} \cong v_j + P_m$ and $F_{1,n} \cong u_i + P_n$ are both subgraphs of $P_{m,n}$ for each vertex $v_j \in V(P_n)$ and $u_i \in V(P_m)$, respectively. By Theorem 3, we have $\beta(F_{1,m}) = \beta(v_j + P_m) = \lfloor \frac{2m+2}{5} \rfloor$ and $\beta(F_{1,n}) = \beta(u_i + P_n) = \lfloor \frac{2n+2}{5} \rfloor$. Let $S_{1,m} = \{s_1, s_2, \ldots, s_\gamma\}$ and $S_{1,n} = \{w_1, w_2, \ldots, w_\alpha\}$ be a metric basis for $v_j + P_m$ and $u_i + P_n$, respectively, for some $v_j \in V(P_n)$ and $u_i \in V(P_m)$ with $\gamma = \lfloor \frac{2m+2}{5} \rfloor$ and $\alpha = \lfloor \frac{2n+2}{5} \rfloor$. By Remark 2, $S_{1,m} \subset V(P_m)$ and $S_{1,n} \subset V(P_n)$. Let $S = S_{1,m} \cup S_{1,n}$. By Remark 1, we can let $S = \{s_1, s_2, \ldots, s_\gamma, w_1, w_2, \ldots, w_\alpha\}$. Observe that

$$r(v_i|S) = (\underbrace{1, \ldots, 1}_{\gamma \ times}, d(v_i, w_1), \ldots, d(v_i, w_\alpha))$$

and $r(v_j|S) = (\underbrace{1, \ldots, 1}_{\gamma \ times}, d(v_j, w_1), \ldots, d(v_j, w_\alpha))$ for every pair of distinct vertices $v_i, v_j \in V(P_n) \backslash S_{1,n}$. Also, we have

$$r(u_i|S) = (d(u_i, s_1), \ldots, d(u_i, s_\gamma), \underbrace{1, \ldots, 1}_{\alpha \ times})$$

and $r(u_j|S) = (d(u_i, s_1), \ldots, d(u_i, s_\gamma), \underbrace{1, \ldots, 1}_{\alpha \ times})$ for every pair of distinct vertices $u_i, u_j \in V(P_m) \backslash S_{1,m}$.

Since $r(u_i|S_{1,m}) \neq r(u_j|S_{1,m})$, we have $r(u_i|S) \neq r(u_j|S)$. Similarly, since $r(v_i|S_{1,n}) \neq r(v_j|S_{1,n})$, we have $r(v_i|S) \neq r(v_j|S)$.

We next show that $r(u_i|S) \neq r(v_j|S)$ for each vertex $u_i \in V(P_m) \backslash S_{1,m}$ and for each vertex $v_j \in V(P_n) \backslash S_{1,n}$. Note that for all $m, n \geq 7$, we have $\gamma, \alpha \geq 3$. If $\gamma = \alpha$, then we are done. Now, suppose $\gamma \neq \alpha$. Without loss of generality, let $\gamma = \alpha + t$ for some positive integer t. Observe that for any vertex $u_i \in V(P_m) \backslash S_{1,m}$ and any vertex $v_j \in V(P_n) \backslash S_{1,n}$, we have

$$r(u_i|S_{1,m}) = (d(u_i, s_1), \ldots, d(u_i, s_\alpha), d(u_i, s_{\alpha+1}), \ldots, d(u_i, s_{\alpha+t}))$$

and $r(v_j|S_{1,m}) = (\underbrace{1, \ldots, 1}_{\alpha \ times}, \underbrace{1, \ldots, 1}_{t \ times})$.

Since $\alpha \geq 3$, $d(u_i, s_k) \neq 1$ for some $1 \leq k \leq \alpha$. Hence, $r(u_i|S_{1,m}) \neq r(v_j|S_{1,m})$ which implies that $r(u_i|S) \neq r(v_j|S)$. It follows that the metric representations of the vertices of $V(P_{m,n}) \backslash S$ with respect to S are distinct. Hence, S is a resolving set for $P_{m,n}$ and it follows that $\beta(P_{m,n}) \leq \beta(F_{1,m}) + \beta(F_{1,n})$.

To prove that $\beta(P_{m,n}) \geq \beta(F_{1,m}) + \beta(F_{1,n})$, it suffices to show that there is no resolving set for $P_{m,n}$ of cardinality less than or equal to $\gamma + \alpha - 1$.

Suppose S' is a resolving set for $P_{m,n}$ of cardinality less than or equal to $\gamma + \alpha - 1$. By Remark 3, we must have $S' = P'_m \cup P'_n$ where P'_m and P'_n are nonempty subsets of $V(P_m)$ and $V(P_n)$, respectively. If $1 \leq |P'_m| < \gamma$, then by Theorem 3, there exist $u_i, u_j \in V(P_m) \backslash P'_m$ such that $r(u_i|P'_m) = r(u_j|P'_m)$. Consequently, we have $r(u_i|S') = r(u_j|S')$, which is a contradiction. If $|P'_m| \geq \gamma$, then $|P'_n| \leq \alpha - 1$. Again, by Theorem 3 there exist $v_i, v_j \in V(P_n)\backslash P'_n$ such that $r(v_i|P'_n) = r(v_j|P'_n)$. Consequently, we have $r(v_i|S') = r(v_j|S')$, which is also a contradiction. If $1 \leq |P'_n| < \alpha$, then by Theorem 3, there exist $v_i, v_j \in V(P_n)\backslash P'_n$ such that $r(v_i|P'_n) = r(v_j|P'_n)$. Consequently, we have $r(v_i|S') = r(v_j|S')$, which is a contradiction. If $|P'_n| \geq \alpha$, then $|P'_m| \leq \gamma - 1$. Again, by Theorem 3 there exist $u_i, u_j \in V(P_m)\backslash P'_m$ such that $r(u_i|P'_m) = r(u_j|P'_m)$. Consequently, we have $r(u_i|S') = r(u_j|S')$, which is also a contradiction. Hence, there exists no resolving set for $P_{m,n}$ of cardinality less than or equal to $\gamma + \alpha - 1$. Therefore, $\beta(P_{m,n}) \geq \beta(F_{1,m}) + \beta(F_{1,n})$ and we conclude that $\beta(P_{m,n}) = \beta(F_{1,m}) + \beta(F_{1,n})$. □

The next theorem is an extension of Theorem 5 to a finite number of paths.

Theorem 6. *Let $m \geq 2$ and $n_i \geq 7$ $(1 \leq i \leq m)$ be positive integers. Then $\beta(P_{n_1} + P_{n_2} + \cdots + P_{n_m}) = \sum_{i=1}^{m} \beta(F_{1,n_i})$.*

Proof. Let $G = P_{n_1} + P_{n_2} + \cdots + P_{n_m}$. Let $S_{1,n_i} = \{s_{i1}, s_{i2}, \ldots, s_{i\alpha_i}\}$ be a metric basis of $u + P_{n_i} \cong F_{1,n_i}$ for each $1 \leq i \leq m$ where $u \notin V(P_{n_i})$. Since $n_i \geq 7$ for each $1 \leq i \leq m$, then $\alpha_i = \lfloor \frac{2n_i+2}{5} \rfloor = \beta(u + P_{n_i}) = \beta(F_{1,n_i})$ by Theorem 3. Let $S = \cup_{i=1}^{m} S_{1,n_i}$. By Remark 2, $S_{1,n_i} \subset V(P_{n_i})$ for each $1 \leq i \leq m$. By Remark 1, we can let $S = \{s_{11}, s_{12}, \ldots, s_{1\alpha_1}, s_{21}, s_{22}, \ldots, s_{2\alpha_2}, \ldots, s_{m1}, s_{m2}, \ldots, s_{m\alpha_m}\}$.

It is not difficult to see that for every pair of distinct vertices $u_{n_i}, u'_{n_i} \in V(P_{n_i})\backslash S_{1,n_i}$, we have $r(u_{n_i}|S_{1,n_i}) \neq r(u'_{n_i}|S_{1,n_i})$. Since $d(u_{n_i}, w) = d(u'_{n_i}, w) = 1$ for any vertex $w \in V(G)\backslash V(P_{n_i})$, it follows that $r(u_{n_i}|S) \neq r(u'_{n_i}|S)$. Following the same arguments as in Theorem 5, it can be shown that $r(u_{n_i}|S) \neq r(u_{n_j}|S)$ for any vertex $u_{n_i} \in V(P_{n_i})\backslash S$ and for any vertex $u_{n_j} \in V(P_{n_j})\backslash S$, $i \neq j$. Hence, the metric representations of the vertices of $V(G)\backslash S$ are distinct. Therefore, S is a resolving set of G. Thus, we have $\beta(G) \leq \sum_{i=1}^{m} \alpha_i$.

To prove that $\beta(G) \geq \sum_{i=1}^{m} \alpha_i$, it suffices to show that there is no resolving set for G of cardinality less than or equal to $(\sum_{i=1}^{m} \alpha_i) - 1$.

Suppose S' is a resolving set for G of cardinality less than or equal to $(\sum_{i=1}^{m} \alpha_i) - 1$. Since $|S'| \leq (\sum_{i=1}^{m} \alpha_i) - 1$, it follows that at least one of the S_{1,n_i}'s, say S_{1,n_1}, is of order less than or equal to $\alpha_1 - 1$. By Theorem 3, there exist $u, v \in V(P_{n_1})\backslash S_{1,n_1}$ such that $r(u|S_{1,n_1}) = r(v|S_{1,n_1})$. Since $d(u, w) = d(v, w) = 1$ for any $w \in V(G)\backslash V(P_{n_1})$, it follows that $r(u|S') = r(v|S')$ contradicting the assumption that S' is a resolving set. Hence, there exists no resolving set of cardinality less than or equal to $(\sum_{i=1}^{m} \alpha_i) - 1$. Therefore, $\beta(G) \geq \sum_{i=1}^{m} \alpha_i$ and we conclude that $\beta(G) = \sum_{i=1}^{m} \beta(F_{1,n_i})$. □

3 The Join of a Finite Number of Cycles

For integers $m, n \geq 3$, let $C_{m,n} = C_m + C_n$ where C_m and C_n are cycles with vertices $u_1, u_2, \ldots, u_m$ and $v_1, v_2, \ldots, v_n$, respectively. We note that the wheels $W_{1,m} \cong v_j + C_m$ for each $v_j \in V(C_n)$ and $W_{1,n} \cong u_i + C_n$ for each $u_i \in V(C_m)$ are subgraphs of $C_{m,n}$. We also note that by symmetry, $C_m + C_n = C_n + C_m$.

Thus, to compute the metric dimension of $C_{m,n}$, it is sufficient to consider the cases when $m = n$ and $m < n$.

Remark 5. If S is a resolving set for $C_{m,n}$, then $S = S_{C_m} \cup S_{C_n}$ where S_{C_m} and S_{C_n} are nonempty subsets of $V(C_m)$ and $V(C_n)$, respectively, with $|S_{C_m}|, |S_{C_n}| \geq 2$. In general, if S is a resolving set for $G = C_{n_1} + C_{n_2} + \cdots + C_{n_m}$ where C_{n_i} is a cycle on $n_i \geq 3$ vertices, then $S = \cup_{i=1}^{m} S_{C_{n_i}}$ where $S_{C_{n_i}}$ is a nonempty subset of $V(C_{n_i})$ for each i with $|S_{C_{n_i}}| \geq 2$.

In the following computations, addition on the subscripts of $u_i \in V(C_m)$ and $v_j \in V(C_n)$ are computed modulo m and n, respectively.

Proposition 3. *If* $m, n \in \{3, 4, 5, 6\}$, *then*

$$\beta(C_{m,n}) = \begin{cases} 4, & \text{if } 3 \leq m \leq 5 \text{ and } 4 \leq n \leq 5, \text{ or } 4 \leq m \leq 5 \text{ and } n = 6 \\ 5, & \text{otherwise.} \end{cases}$$

Proof. We consider four cases.

CASE 1. $m = 3$ and $n = 3$

Note that $C_{3,3} \cong K_6$. Therefore, $\beta(C_{3,3}) = 5$ by Theorem 1.

CASE 2. $3 \leq m \leq 5$ and $4 \leq n \leq 5$, or $4 \leq m \leq 5$ and $n = 6$

It is not difficult to see that the set $S = \{u_1, u_2, v_1, v_2\}$ is a resolving set for $C_{m,n}$ ($3 \leq m \leq 5$ and $4 \leq n \leq 5$) and the set $S = \{u_1, u_2, v_1, v_3\}$ is a resolving set for $C_{m,n}$ ($4 \leq m \leq 5$ and $n = 6$). Thus, we have $\beta(C_{m,n}) \leq 4$. By Remark 5, we have $\beta(C_{m,n}) \geq 4$. Therefore, we conclude that $\beta(C_{m,n}) = 4$.

CASE 3. $m = 3$ and $n = 6$

Let $S = \{u_1, u_2, v_1, v_2, v_3\}$. Then we have the following:
$r(u_3|S) = (1, 1, 1, 1, 1)$, $r(v_4|S) = (1, 1, 2, 2, 1)$, $r(v_5|S) = (1, 1, 2, 2, 2)$ and $r(v_6|S) = (1, 1, 1, 2, 2)$. Clearly, these representations are distinct. Thus, S is a resolving set and so $\beta(C_{3,6}) \leq 5$.

To prove that $\beta(C_{3,6}) \geq 5$, it suffices to show that there exists no resolving set for $C_{3,6}$ of cardinality 4 by Remark 5.

Suppose S' is a resolving set for $C_{3,6}$ of cardinality 4. By Remark 5, we must have $S' = \{u_i, u_j\} \cup \{v_k, v_l\}$ for some $u_i, u_j \in V(C_3)$ and for some $v_k, v_l \in V(C_6)$. By Remark 1, we can let $S' = \{u_i, u_j, v_k, v_l\}$. Also, we can assume that $k \bmod 6 < l \bmod 6$. We then have the following possibilities:

(1) If $d(v_k, v_l) = 1$, then $r(v_{k+3}|S') = r(v_{k+4}|S')$ or $r(v_{k+2}|S') = r(v_{k+3}|S')$.

(2) If $d(v_k, v_l) = 2$, then $r(u_x|S') = r(v_{k+1}|S')$ or $r(u_x|S') = r(v_{k-1}|S')$, where $u_x \in V(C_3)\backslash\{u_i, u_j\}$.

(3) If $d(v_k, v_l) = 3$, then $r(v_{k+2}|S') = r(v_{k+4}|S')$ and $r(v_{k+1}|S') = r(v_{k-1}|S')$.

In any case, we arrive at a contradiction. This implies that there exists no resolving set for $C_{3,6}$ of cardinality 4. Thus, we have $\beta(C_{3,6}) \geq 5$. Therefore, we conclude that $\beta(C_{3,6}) = 5$.

CASE 4. $m = 6$ and $n = 6$

Let $S = \{u_1, u_2, u_3, v_1, v_3\}$. Then we have the following:
$r(u_4|S) = (2, 2, 1, 1, 1)$, $r(u_5|S) = (2, 2, 2, 1, 1)$, $r(u_6|S) = (1, 2, 2, 1, 1)$,

$r(v_2|S) = (1,1,1,1,1)$, $r(v_4|S) = (1,1,1,2,1)$, $r(v_5|S) = (1,1,1,2,2)$ and $r(v_6|S) = (1,1,1,1,2)$. Clearly, these representations are distinct. Thus, S is a resolving set and so $\beta(C_{6,6}) \leq 5$.

To prove that $\beta(C_{6,6}) \geq 5$, it suffices to show that there exists no resolving set for $C_{6,6}$ of cardinality 4.

Suppose S' is a resolving set for $C_{6,6}$ of cardinality 4. By Remark 5, we must have $S' = \{u_i, u_j\} \cup \{v_k, v_l\}$ for some $u_i, u_j \in V(C_6)$ and for some $v_k, v_l \in V(C_6)$. By Remark 1, we can let $S' = \{u_i, u_j, v_k, v_l\}$. Also, we can assume that $i \bmod 6 < j \bmod 6$ and $k \bmod 6 < l \bmod 6$. It is sufficient to consider the following possibilities:

(1) If $d(u_i, u_j) = 1$, then $r(u_{i+3}|S') = r(u_{i+4}|S')$ or $r(u_{i+2}|S') = r(u_{i+3}|S')$. If $d(v_k, v_l) = 1$, then $r(v_{k+3}|S') = r(v_{k+4}|S')$ or $r(v_{k+2}|S') = r(v_{k+3}|S')$.

(2) If $d(u_i, u_j) = 2$ and $d(v_k, v_l) = 2$, then $r(u_{i+1}|S') = r(v_{k+1}|S')$ or $r(u_{i-1}|S') = r(v_{k-1}|S')$ or $r(u_{i+1}|S') = r(v_{k-1}|S')$ or $r(u_{i-1}|S') = r(v_{k+1}|S')$.

(3) If $d(u_i, u_j) = 3$, then $r(u_{i+2}|S') = r(u_{i+4}|S')$ and $r(u_{i+1}|S') = r(u_{i-1}|S')$. If $d(v_k, v_l) = 3$, then $r(v_{k+2}|S') = r(v_{k+4}|S')$ and $r(v_{k+1}|S') = r(v_{k-1}|S')$.

In any case, we arrive at a contradiction. This implies that there exists no resolving set for $C_{6,6}$ of cardinality 4. Thus, we have $\beta(C_{6,6}) \geq 5$. Therefore, we conclude that $\beta(C_{6,6}) = 5$. □

Proposition 4. *Let $m \in \{3,4,5,6\}$ and $n \geq 7$ be an integer. Then $\beta(C_{m,n}) = 2 + \beta(W_{1,n})$.*

Proof. As noted above, $W_{1,n} \cong u_i + C_n$ for each vertex $u_i \in V(C_m)$. By Theorem 2, we have $\beta(W_{1,n}) = \beta(u_i + C_n) = \lfloor \frac{2n+2}{5} \rfloor$ for some vertex $u_i \in V(C_m)$. Let $S_{1,n} = \{w_1, w_2, \ldots, w_\alpha\}$ be a metric basis for $u_i + C_n$ where $\alpha = \lfloor \frac{2n+2}{5} \rfloor$. By Remark 2, $S_{1,n}$ is a subset of $V(C_n)$. It follows that $r(v_i|S_{1,n}) \neq r(v_j|S_{1,n})$ for every pair of distinct vertices $v_i, v_j \in V(C_n)\backslash S_{1,n}$. Note that $r(v_j|S_{1,n})$ consists of σ 1's and $(\alpha - \sigma)$ 2's where $0 \leq \sigma \leq 2$ for each $v_j \in V(C_n)\backslash S_{1,n}$.

Consider first the case when $m = 3, 4$ and 5. Let $S = \{u_1, u_2\} \cup S_{1,n}$. By Remark 1, we can let $S = \{u_1, u_2, w_1, w_2, \ldots, w_\alpha\}$. If $m = 3$, we have $r(u_3|S) = (1,1,\underbrace{1,\ldots,1}_{\alpha\ times})$. If $m = 4$, then we have $r(u_3|S) = (2,1,\underbrace{1,\ldots,1}_{\alpha\ times})$ and $r(u_4|S) = (1,2,\underbrace{1,\ldots,1}_{\alpha\ times})$. If $m = 5$, then we have $r(u_3|S) = (2,1,\underbrace{1,\ldots,1}_{\alpha\ times})$, $r(u_4|S) = (2,2,\underbrace{1,\ldots,1}_{\alpha\ times})$ and $r(u_5|S) = (1,2,\underbrace{1,\ldots,1}_{\alpha\ times})$. Observe that for each distinct vertices $v_i, v_j \in V(C_n)\backslash S_{1,n}$, we have

$$r(v_i|S) = (1,1,d(v_i,w_1),\ldots,d(v_i,w_\alpha))$$

and $r(v_j|S) = (1,1,d(v_j,w_1),\ldots,d(v_j,w_\alpha))$.

Since $r(v_i|S_{1,n}) \neq r(v_j|S_{1,n})$, it follows that $r(v_i|S) \neq r(v_j|S)$. Also, since $r(v_j|S_{1,n})$ consists of σ 1's and $(\alpha - \sigma)$ 2's where $0 \leq \sigma \leq 2$ for each $v_j \in V(C_n)\backslash S_{1,n}$, we have $r(u_i|S) \neq r(v_j|S)$ for each $u_i \in V(C_m)\backslash S$ and $v_j \in V(C_n)\backslash S$. Hence, the metric representations of the vertices of $V(C_{m,n})\backslash S$

with respect to S are distinct. Therefore, S is a resolving set and we conclude that $\beta(C_{m,n}) \leq 2 + \alpha$.

To prove that $\beta(C_{m,n}) \geq 2 + \alpha$, it suffices to show that there exists no resolving set of cardinality less than or equal to $1 + \alpha$.

Suppose S' is a resolving set for $C_{m,n}$ of cardinality less than or equal to $1+\alpha$. By Remark 5, we must have $S' = C'_m \cup C'_n$ where C'_m and C'_n are nonempty subsets of $V(C_m)$ and $V(C_n)$, respectively, with $|C'_m| \geq 2$ and $|C'_n| \geq 2$. It follows that $|C'_n| \leq \alpha - 1$. By Theorem 2, there exist $v_i, v_j \in V(C_n)\backslash C'_n$ such that $r(v_i|C'_n) = r(v_j|C'_n)$. Since $d(u_i, v_j) = 1$ for all j $(1 \leq j \leq n)$, it follows that $r(v_i|S') = r(v_j|S')$ which is a contradiction. Therefore, there exists no resolving set of cardinality less than or equal to $1 + \alpha$. Hence, we have $\beta(C_{m,n}) \geq 2 + \alpha$. Therefore, we have $\beta(C_{m,n}) = 2 + \beta(W_{1,n})$.

Following the same arguments above, we see that the set $S = \{u_1, u_3\} \cup S_{1,n}$ is a metric basis for $C_{6,n}$. Hence, we conclude that $\beta(C_{m,n}) = 2 + \beta(W_{1,n})$. □

Theorem 7. *Let $m, n \geq 7$ be integers. Then $\beta(C_{m,n}) = \beta(W_{1,m}) + \beta(W_{1,n})$.*

Proof. Again, we note that the wheels $W_{1,m} \cong v_j + C_m$ and $W_{1,n} \cong u_i + C_n$ are both subgraphs of $C_{m,n}$ for each vertex $v_j \in V(C_n)$ and for each vertex $u_i \in V(C_m)$. By Theorem 2, we have $\beta(W_{1,m}) = \beta(v_j + C_m) = \lfloor \frac{2m+2}{5} \rfloor$ and $\beta(W_{1,n}) = \beta(u_i + C_n) = \lfloor \frac{2n+2}{5} \rfloor$. Let $S_{1,m} = \{s_1, s_2, \ldots, s_\gamma\}$ be a metric basis for $v_j + C_m$ and $S_{1,n} = \{w_1, w_2, \ldots, w_\alpha\}$ be a metric basis for $u_i + C_n$, respectively, with $\gamma = \lfloor \frac{2m+2}{5} \rfloor$ and $\alpha = \lfloor \frac{2n+2}{5} \rfloor$. By Remark 2, $S_{1,m} \subset V(C_m)$ and $S_{1,n} \subset V(C_n)$. It follows that $r(u_i|S_{1,m}) \neq r(u_j|S_{1,m})$ for every pair of distinct vertices $u_i, u_j \in V(C_m)\backslash S_{1,m}$ and $r(v_i|S_{1,n}) \neq r(v_j|S_{1,n})$ for every pair of distinct vertices $v_i, v_j \in V(C_n)\backslash S_{1,n}$. Let $S = S_{1,m} \cup S_{1,n}$. Again, by Remark 1, we can let $S = \{s_1, \ldots, s_\gamma, w_1, \ldots, w_\alpha\}$. Observe that $r(v_i|S) = (\underbrace{1, \ldots, 1}_{\gamma \text{ times}}, d(v_i, w_1), \ldots, d(v_i, w_\alpha))$ and $r(v_j|S) = (\underbrace{1, \ldots, 1}_{\gamma \text{ times}}, d(v_j, w_1), \ldots, d(v_j, w_\alpha))$ for every pair of distinct vertices $v_i, v_j \in V(C_n)\backslash S_{1,n}$. Also, we have

$$r(u_i|S) = (d(u_i, s_1), \ldots, d(u_i, s_\gamma), \underbrace{1, \ldots, 1}_{\alpha \text{ times}})$$

and $r(u_j|S) = (d(u_i, s_1), \ldots, d(u_i, s_\gamma), \underbrace{1, \ldots, 1}_{\alpha \text{ times}})$ for every pair of distinct vertices $u_i, u_j \in V(C_m)\backslash S_{1,m}$.

Since $r(u_i|S_{1,m}) \neq r(u_j|S_{1,m})$, we have $r(u_i|S) \neq r(u_j|S)$. Similarly, since $r(v_i|S_{1,n}) \neq r(v_j|S_{1,n})$, we have $r(v_i|S) \neq r(v_j|S)$.

We next show that $r(u_i|S) \neq r(v_j|S)$ for each vertex $u_i \in V(C_m)\backslash S_{1,m}$ and for each vertex $v_j \in V(C_n)\backslash S_{1,n}$. Since $m, n \geq 7$, we have $\gamma, \alpha \geq 3$. If $\gamma = \alpha$, then we are done. Now, suppose $\gamma \neq \alpha$. Without loss of generality, let $\gamma = \alpha + t$ for some positive integer t. Observe that for any vertex $u_i \in V(C_m)\backslash S_{1,m}$ and any vertex $v_j \in V(C_n)\backslash S_{1,n}$, we have

$$r(u_i|S_{1,m}) = (d(u_i, s_1), \ldots, d(u_i, s_\alpha), d(u_i, s_{\alpha+1}), \ldots, d(u_i, s_{\alpha+t}))$$

and $r(v_j|S_{1,m}) = (\underbrace{1, \ldots, 1}_{\alpha\ times}, \underbrace{1, \ldots, 1}_{t\ times})$.

Since $\alpha \geq 3$, $d(u_i, s_k) \neq 1$ for some k $(1 \leq k \leq \alpha)$. Hence, $r(u_i|S_{1,m}) \neq r(v_j|S_{1,m})$, which implies that $r(u_i|S) \neq r(v_j|S)$. It follows that the metric representations of the vertices of $V(C_{m,n})\backslash S$ with respect to S are distinct. Hence, S is a resolving set for $C_{m,n}$ and it follows that $\beta(C_{m,n}) \leq \beta(W_{1,m}) + \beta(W_{1,n})$.

To prove that $\beta(C_{m,n}) \geq \beta(W_{1,m}) + \beta(W_{1,n})$, it suffices to show that there is no resolving set for $C_{m,n}$ of cardinality less than or equal to $\gamma + \alpha - 1$.

Suppose S' is a resolving set of cardinality less than or equal to $\gamma+\alpha-1$. By Remark 5, we must have $S' = C'_m \cup C'_n$, where C'_m and C'_n are nonempty subsets of $V(C_m)$ and $V(C_n)$, respectively, with $|C'_m| \geq 2$ and $|C'_n| \geq 2$. If $2 \leq |C'_m| < \gamma$, then by Theorem 2, there exist $u_i, u_j \in V(C_m)\backslash C'_m$ such that $r(u_i|C'_m) = r(u_j|C'_m)$. Consequently, we have $r(u_i|S') = r(u_j|S')$, which is a contradiction. If $|C'_m| \geq \gamma$, then $|C'_n| \leq \alpha-1$. Again, by Theorem 2 there exist $v_i, v_j \in V(C_n)\backslash C'_n$ such that $r(v_i|C'_n) = r(v_j|C'_n)$. Consequently, we have $r(v_i|S') = r(v_j|S')$, which is also a contradiction. If $2 \leq |C'_n| < \alpha$, then by Theorem 2, there exist $v_i, v_j \in V(C_n)\backslash C'_n$ such that $r(v_i|C'_n) = r(v_j|C'_n)$. Consequently, we have $r(v_i|S') = r(v_j|S')$, which is a contradiction. If $|C'_n| \geq \alpha$, then $|C'_m| \leq \gamma - 1$. Again, by Theorem 2 there exist $u_i, u_j \in V(C_m)\backslash C'_m$ such that $r(u_i|C'_m) = r(u_j|C'_m)$. Consequently, we have $r(u_i|S') = r(u_j|S')$ which is also a contradiction. Hence, there exists no resolving set of cardinality less than or equal to $\gamma + \alpha - 1$. Therefore, $\beta(C_{m,n}) \geq \beta(W_{1,m}) + \beta(W_{1,n})$ and we conclude that $\beta(C_{m,n}) = \beta(W_{1,m}) + \beta(W_{1,n})$. □

The next theorem is an extension of Theorem 7 to a finite number of cycles.

Theorem 8. *Let $m \geq 2$ and $n_i \geq 7$ $(1 \leq i \leq m)$ be integers. Then $\beta(C_{n_1} + C_{n_2} + \cdots + C_{n_m}) = \sum_{i=1}^{m} \beta(W_{1,n_i})$.*

Proof. Let $G = C_{n_1} + C_{n_2} + \cdots + C_{n_m}$. Let $S_{1,n_i} = \{s_{i1}, s_{i2}, \ldots, s_{i\alpha_i}\}$ be a metric basis of $u + C_{n_i} \cong W_{1,n_i}$ for each $1 \leq i \leq m$ where $u \notin V(C_{n_i})$. Since $n_i \geq 7$ for each $1 \leq i \leq m$, then $\alpha_i = \lfloor \frac{2n_i+2}{5} \rfloor = \beta(u + C_{n_i}) = \beta(W_{1,n_i})$ by Theorem 2. Let $S = \cup_{i=1}^{m} S_{1,n_i}$. By Remark 2, $S_{1,n_i} \subset V(C_{n_i})$. By Remark 1, we can let $S = \{s_{11}, s_{12}, \ldots, s_{1\alpha_1}, s_{21}, s_{22}, \ldots, s_{2\alpha_2}, \ldots, s_{m1}, s_{m2}, \ldots, s_{m\alpha_m}\}$.

It is not difficult to see that for every pair of distinct vertices $u_{n_i}, u'_{n_i} \in V(C_{n_i})\backslash S_{1,n_i}$, we have $r(u_{n_i}|S_{1,n_i}) \neq r(u'_{n_i}|S_{1,n_i})$. Since $d(u_{n_i}, w) = d(u'_{n_i}, w) = 1$ for any vertex $w \in V(G)\backslash V(C_{n_i})$, it follows that $r(u_{n_i}|S) \neq r(u'_{n_i}|S)$. Following the same arguments as in Theorem 7, it can be shown that $r(u_{n_i}|S) \neq r(u_{n_j}|S)$ for any vertex $u_{n_i} \in V(C_{n_i})\backslash S$ and for any vertex $u_{n_j} \in V(C_{n_j})\backslash S$, $i \neq j$. Hence, the metric representations of the vertices of $V(G)\backslash S$ are distinct. Therefore, S is a resolving set of G. Thus, we have $\beta(G) \leq \sum_{i=1}^{m} \alpha_i$.

To prove that $\beta(G) \geq \sum_{i=1}^{m} \alpha_i$, it suffices to show that there is no resolving set for G of cardinality less than or equal to $(\sum_{i=1}^{m} \alpha_i) - 1$.

Suppose S' is a resolving set of cardinality less than or equal to $(\sum_{i=1}^{m} \alpha_i)-1$. Since $|S'| \leq (\sum_{i=1}^{m} \alpha_i)-1$, it follows that at least one of the $S_{1,n_i}{}'s$, say S_{1,n_1}, is of order less than or equal to $\alpha_1 - 1$. By Theorem 2, there exist $u, v \in V(C_{n_1})\backslash S_{1,n_1}$ such that $r(u|S_{1,n_1}) = r(v|S_{1,n_1})$. Since $d(u,w) = d(v,w) = 1$ for any $w \in V(G)\backslash V(C_{n_1})$, it follows that $r(u|S') = r(v|S')$, which is a contradiction. Hence, there exists no resolving set of cardinality less than or equal to $(\sum_{i=1}^{m} \alpha_i) - 1$. Therefore, $\beta(G) \geq \sum_{i=1}^{m} \alpha_i$ and we conclude that $\beta(G) = \sum_{i=1}^{m} \beta(W_{1,n_i})$. □

4 The Join of a Path and a Cycle

In this section, we compute the metric dimension of $P_m + C_n$, where P_m is the path on $m \geq 2$ vertices and C_n is the cycle on $n \geq 3$ vertices. We note that the fan $F_{1,m} \cong v_j + P_m$ for each $v_j \in V(C_n)$ and the wheel $W_{1,n} \cong u_i + C_n$ for each $u_i \in V(P_m)$ are subgraphs of $P_m + C_n$. Recall that if S is a resolving set for $G + H$, then $S = S_G \cup S_H$, where S_G and S_H are nonempty subsets of $V(G)$ and $V(H)$, respectively, by Remark 3.

Remark 6. If S is a resolving set for $P_m + C_n$, then $S = S_{P_m} \cup S_{C_n}$ where $S_{P_m} \subset V(P_m)$ and $S_{C_n} \subset V(C_n)$ with $|S_{P_m}| \geq 1$ and $|S_{C_n}| \geq 2$. In particular, $|S_{P_m}| = 1$ if $m = 2, 3$, and $|S_{P_m}| \geq 2$ if $m \geq 4$.

Proposition 5. *If $m \in \{2,3,4,5,6\}$ and $n \in \{3,4,5,6\}$, then*

$$\beta(P_m + C_n) = \begin{cases} 3, \textit{ if } m = 2,3 \textit{ and } n = 4,5 \\ 5, \textit{ if } m = 6 \textit{ or } n = 3,6 \\ 4, \textit{ otherwise.} \end{cases}$$

Proof. We consider three cases.

CASE 1. $m = 2,3$ and $n = 4,5$

One can easily check that the set $S = \{u_1, v_1, v_2\}$ is a resolving set for P_m+C_n $(m = 2,3$ and $n = 4,5)$. Therefore, by Remark 6, we have $\beta(P_m + C_n) = 3$.

CASE 2. $m = 2,3$ and $n = 3,6$, or $m = 4,5$ and $n = 3,4,5,6$, or $m = 6$ and $n = 4,5$.

Note that $P_2 + C_3 \cong K_5$. Therefore, $\beta(P_2 + C_3) = 4$ by Theorem 1.

It is not difficult to verify that the set $S = \{u_1, v_1, v_2, v_3\}$ is a resolving set for both P_2+C_6, P_3+C_3 and P_3+C_6. It follows that $\beta(P_2+C_6) = \beta(P_3+C_3) = \beta(P_3 + C_6) \leq 4$. To prove the reverse inequality, it suffices to prove that there exists no resolving set for $P_2 + C_6$, $P_3 + C_3$ and $P_3 + C_6$ of cardinality 3 by Remark 6.

Suppose S' is a resolving for $P_2 + C_6$ of cardinality 3. Then, $S' = \{u_i\} \cup \{v_j, v_k\}$ for some $u_i \in V(P_2)$ and for some $v_j, v_k \in V(C_6)$ by Remark 6. It is not difficult the check that there exist two vertices in $V(C_6)\backslash\{v_j, v_k\}$ or a vertex in $V(C_6)\backslash\{v_j, v_k\}$ and the vertex in $V(P_2)\backslash\{u_i\}$ with the same metric representation with respect to S', contradicting the assumption that S' is a resolving set. Thus, $\beta(P_2+C_6) \geq 4$. Therefore, we conclude that $\beta(P_2+C_6) = 4$.

Now, suppose S' is a resolving for $P_3 + C_n$ $(n = 3,6)$ of cardinality 3. Then, $S' = \{u_i\} \cup \{v_j, v_k\}$, for some $u_i \in V(P_3)$ and for some $v_j, v_k \in V(C_n)$

by Remark 6. It is not difficult to check that there exist two vertices in $V(C_n)\backslash\{v_j, v_k\}$, or a vertex in $V(C_n)\backslash\{v_j, v_k\}$ and a vertex in $V(P_3)\backslash\{u_i\}$, or two vertices in $V(P_3)\backslash\{u_i\}$ with the same metric representation with respect to S', contradicting the assumption that S' is a resolving set. Thus, $\beta(P_3+C_3) \geq 4$ and $\beta(P_3 + C_6) \geq 4$. Therefore, we conclude that $\beta(P_3 + C_3) = \beta(P_3 + C_6) = 4$.

If $m = 4, 5$ and $n = 3, 4, 5$, the set $S = \{u_1, u_m, v_1, v_2\}$ is a resolving set for $P_m + C_n$ and hence $\beta(P_m + C_n) \leq 4$. By Remark 6, $\beta(P_m + C_n) \geq 4$. Therefore, $\beta(P_m + C_n) = 4$.

It is not difficult to verify that the set $S = \{u_1, u_m, v_1, v_3\}$ is a resolving set for $P_m + C_n$, where $m = 4, 5$ and $n = 6$. It follows that $\beta(P_m + C_n) \leq 4$. By Remark 6, $\beta(P_m + C_n) \geq 4$. Therefore, $\beta(P_m + C_n) = 4$.

For the case when $m = 6$ and $n = 4, 5$, the set $S = \{u_2, u_4, v_1, v_2\}$ is a resolving set for $P_m + C_n$. It follows that $\beta(P_m + C_n) \leq 4$. By Remark 6, $\beta(P_m + C_n) \geq 4$. Therefore, $\beta(P_m + C_n) = 4$.

Case 3. $m = 6$ and $n = 3, 6$

It is not difficult to verify that the set $S = \{u_2, u_4, v_1, v_2, v_3\}$ is a resolving set for $P_6 + C_n$ $(n = 3, 6)$. It follows that $\beta(P_6 + C_n) \leq 5$. To prove the reverse inequality, we show that there exists no resolving set for $P_6 + C_n$ of cardinality 4 by Remark 6.

Suppose S' is a resolving for $P_6 + C_n$ of cardinality 4. Then, $S' = \{u_i, u_j\} \cup \{v_k, v_l\}$ for some $u_i, u_j \in V(P_6)$ and for some $v_k, v_l \in V(C_n)$ by Remark 6. It is not difficult to check that there exist two vertices in $V(P_m)\backslash\{u_i, u_j\}$, or two vertices in $V(C_n)\backslash\{v_k, v_l\}$, or a vertex in $V(P_m)\backslash\{u_i, u_j\}$ and a vertex in $V(C_n)\backslash\{v_k, v_l\}$ with the same metric representation with respect to S', contradicting the assumption that S' is a resolving set. Thus, there exists no resolving set for $P_6 + C_n$ of order 4. Therefore, $\beta(P_6 + C_n) \geq 5$ and we conclude that $\beta(P_6 + C_n) = 5$. □

Proposition 6. *Let $m \in \{2, 3, 4, 5, 6\}$ and $n \geq 7$ be an integer. Then*

$$\beta(P_m + C_n) = \begin{cases} 1 + \beta(W_{1,n}), \text{ if } m = 2, 3 \\ 2 + \beta(W_{1,n}), \text{ if } m = 4, 5, 6. \end{cases}$$

Proof. As noted above, $W_{1,n} \cong u_i + C_n$ for each vertex $u_i \in V(P_m)$. By Theorem 2, we have $\beta(W_{1,n}) = \beta(u_i + C_n) = \lfloor \frac{2n+2}{5} \rfloor$ for some vertex $u_i \in V(P_m)$. Let $S_{1,n} = \{w_1, w_2, \ldots, w_\alpha\}$ be a metric basis for $u_i + C_n$ where $\alpha = \lfloor \frac{2n+2}{5} \rfloor$. By Remark 2, $S_{1,n}$ is a subset of $V(C_n)$. It follows that $r(v_i|S_{1,n}) \neq r(v_j|S_{1,n})$ for every pair of distinct vertices $v_i, v_j \in V(C_n)\backslash S_{1,n}$. Note that $r(v_j|S_{1,n})$ consists of σ 1's and $(\alpha - \sigma)$ 2's where $0 \leq \sigma \leq 2$ for each $v_j \in V(C_n)\backslash S_{1,n}$.

Consider first the case when $m = 2$ and 3. Let $S = \{u_1\} \cup S_{1,n}$. By Remark 1, we can let $S = \{u_1, w_1, w_2, \ldots, w_\alpha\}$. If $m = 2$, then we have $r(u_2|S) = (1, \underbrace{1, \ldots, 1}_{\alpha \text{ times}})$. If $m = 3$, then we have $r(u_2|S) = (1, \underbrace{1, \ldots, 1}_{\alpha \text{ times}})$ and $r(u_3|S) = (2, \underbrace{1, \ldots, 1}_{\alpha \text{ times}})$. Observe that for each distinct vertices $v_i, v_j \in V(C_n)\backslash S_{1,n}$, we have $r(v_i|S) = (1, d(v_i, w_1), \ldots, d(v_i, w_\alpha))$ and $r(v_j|S) = (1, d(v_j, w_1), \ldots, d(v_j, w_\alpha))$.

Since $r(v_i|S_{1,n}) \neq r(v_j|S_{1,n})$, it follows that $r(v_i|S) \neq r(v_j|S)$. Also, since $r(v_j|S_{1,n})$ consists of σ 1's and $(\alpha - \sigma)$ 2's where $0 \leq \sigma \leq 2$ for each $v_j \in V(C_n)\backslash S_{1,n}$, we have $r(u_i|S) \neq r(v_j|S)$ for each $u_i \in V(C_n)\backslash S$ and $v_j \in V(C_n)\backslash S$. Hence, the metric representations of the vertices of $V(P_m+C_n)\backslash S$ with respect to S are distinct. Therefore, S is a resolving set and we conclude that $\beta(P_m + C_n) \leq 1 + \alpha$.

To prove that $\beta(P_m + C_n) \geq 1 + \alpha$, it suffices to show that there exists no resolving set of cardinality less than or equal to α.

Suppose S' is a resolving set $P_m + C_n$ of cardinality less than or equal to α. By Remark 6, we must have $S' = P'_m \cup C'_n$ where P'_m and C'_n are nonempty subsets of $V(P_m)$ and $V(C_n)$, respectively, with $|P'_m| \geq 1$ and $|C'_n| \geq 2$. It follows that $|C'_n| \leq \alpha - 1$. By Theorem 2, there exist $v_i, v_j \in V(C_n)\backslash C'_n$ such that $r(v_i|C'_n) = r(v_j|C'_n)$. Since $d(u_i, v_j) = 1$ for all j $(1 \leq j \leq n)$, it follows that $r(v_i|S') = r(v_j|S')$, which is a contradiction. Therefore, there exists no resolving set of cardinality less than or equal to α. Hence, we have $\beta(P_m + C_n) \geq 1 + \alpha$. Therefore, we have $\beta(P_m + C_n) = 1 + \beta(W_{1,n})$.

Following the same arguments above, we see that the set $S = \{u_2, u_4\} \cup S_{1,n}$ is a metric basis for $P_m + C_n$ where $m = 4, 5, 6$. Hence, we conclude that $\beta(P_m + C_n) = 2 + \beta(W_{1,n})$. □

Proposition 7. *Let $n \in \{3, 4, 5, 6\}$ and $m \geq 7$ be an integer. Then $\beta(P_m + C_n) = \beta(F_{1,m}) + 2$.*

Proof. As noted above, $F_{1,m} \cong v_j + P_m \subset P_m + C_n$ for each vertex $v_j \in V(C_n)$. By Theorem 3, we have $\beta(F_{1,m}) = \beta(v_j + P_m) = \lfloor \frac{2m+2}{5} \rfloor$ for some vertex $v_j \in V(C_n)$. Let $S_{1,m} = \{w_1, w_2, \ldots, w_\alpha\}$ be a metric basis for $v_j + P_m$ where $\alpha = \lfloor \frac{2m+2}{5} \rfloor$. By Remark 2, $S_{1,m}$ is a subset of $V(P_m)$. It follows that $r(u_i|S_{1,m}) \neq r(u_j|S_{1,m})$ for every pair of distinct vertices $u_i, u_j \in V(P_m)\backslash S_{1,m}$. Note that $r(u_j|S_{1,m})$ consists of σ 1's and $(\alpha - \sigma)$ 2's where $0 \leq \sigma \leq 2$ for each $u_j \in V(P_m)\backslash S_{1,m}$.

Consider first the case when $n = 3$. Let $S = \{v_1, v_2\} \cup S_{1,m}$. By Remark 1, we can let $S = \{v_1, v_2, w_1, w_2, \ldots, w_\alpha\}$. Then we have $r(v_3|S) = (1, 1, \underbrace{1, \ldots, 1}_{\alpha \text{ times}})$.

Observe that for each distinct vertices $u_i, u_j \in V(P_m)\backslash S_{1,m}$, we have $r(u_i|S) = (1, 1, d(u_i, w_1), \ldots, d(u_i, w_\alpha))$ and $r(u_j|S) = (1, 1, d(u_j, w_1), \ldots, d(u_j, w_\alpha))$.

Since $r(u_i|S_{1,m}) \neq r(u_j|S_{1,m})$, it follows that $r(u_i|S) \neq r(u_j|S)$. Also, since $r(u_j|S_{1,m})$ consists of σ 1's and $(\alpha - \sigma)$ 2's where $0 \leq \sigma \leq 2$ for each $u_j \in V(P_m)\backslash S_{1,m}$, we have $r(v_3|S) \neq r(u_j|S)$ for each $u_j \in V(P_m)\backslash S_{1,m}$. Hence, the metric representations of the vertices of $V(P_m + C_3)\backslash S$ with respect to S are distinct. Therefore, S is a resolving set and we conclude that $\beta(P_m+C_3) \leq \alpha+2$.

To prove that $\beta(P_m + C_3) \geq 2 + \alpha$, it suffices to show that there exists no resolving set of cardinality less than or equal to $\alpha + 1 = \beta(F_{1,m}) + 1$.

Suppose S' is a resolving set for $P_m + C_3$ of cardinality less than or equal to $\alpha+1$. By Remark 3, $S' = S_{C_3} \cup S_{P_m}$ where S_{C_3} and S_{P_m} are nonempty subsets of $V(C_3)$ and $V(P_m)$, respectively. Since $|S_{C_3}| \geq 2$, then $|S_{P_m}| \leq \beta(F_{1,m}) - 1$. By Theorem 3, there exist $u_i, u_j \in V(P_m)\backslash S_{P_m}$ such that $r(u_i|S_{P_m}) = r(u_j|S_{P_m})$.

Consequently, $r(u_i|S') = r(u_j|S')$, which is a contradiction. Hence, there exists no resolving set for $P_m + C_3$ of cardinality less than or equal to $\beta(F_{1,m}) + 1$. Therefore, $\beta(P_m + C_3) \geq \beta(F_{1,m}) + 2$ and we conclude that $\beta(P_m + C_3) = \beta(F_{1,m}) + 2$.

Following the same arguments above, we see that the set $S = \{v_1, v_2\} \cup S_{1,m}$ is also a metric basis for $P_m + C_4$ and $P_m + C_5$. Hence, we conclude that $\beta(P_m + C_4) = \beta(P_m + C_5) = \beta(F_{1,m}) + 2$. Furthermore, the set $S = \{v_1, v_3\} \cup S_{1,m}$ is a metric basis for $P_m + C_6$. Hence, we conclude that $\beta(P_m + C_6) = \beta(F_{1,m}) + 2$. □

Theorem 9. *Let $m, n \geq 7$ be integers. Then $\beta(P_m + C_n) = \beta(F_{1,m}) + \beta(W_{1,n})$.*

Proof. Recall that for each vertex $u_i \in V(P_m)$ and for each vertex $v_j \in V(C_n)$, the wheel graph $W_{1,n} \cong u_i + C_n$ and the fan graph $F_{1,m} \cong v_j + P_m$, respectively, are subgraphs of $P_m + C_n$.

Let $S_{1,m} = \{s_1, s_2, \ldots, s_\gamma\}$ and $S_{1,n} = \{w_1, w_2, \ldots, w_\alpha\}$ be a metric basis for $F_{1,m} \cong v_j + P_m$ for some vertex $v_j \in V(C_n)$ and $W_{1,n} \cong u_i + C_n$ for some vertex $u_i \in V(P_m)$, respectively, with $\gamma = \beta(F_{1,m})$ and $\alpha = \beta(W_{1,n})$. By Remark 2, $S_{1,m} \subset V(P_m)$ and $S_{1,n} \subset V(C_n)$. Let $S = S_{1,m} \cup S_{1,n}$. By Remark 1, we can let $S = \{s_1, s_2, \ldots, s_\gamma, w_1, w_2, \ldots, w_\alpha\}$. Observe that

$$r(v_i|S) = (\underbrace{1, \ldots, 1}_{\gamma \text{ times}}, d(v_i, w_1), \ldots, d(v_i, w_\alpha))$$

and $r(v_j|S) = (\underbrace{1, \ldots, 1}_{\gamma \text{ times}}, d(v_j, w_1), \ldots, d(v_j, w_\alpha))$ for every pair of distinct vertices $v_i, v_j \in V(C_n) \backslash S_{1,n}$. Also, we have $r(u_i|S) = (d(u_i, s_1), \ldots, d(u_i, s_\gamma), \underbrace{1, \ldots, 1}_{\alpha \text{ times}})$ and $r(u_j|S) = (d(u_i, s_1), \ldots, d(u_i, s_\gamma), \underbrace{1, \ldots, 1}_{\alpha \text{ times}})$ for every pair of distinct vertices $u_i, u_j \in V(P_m) \backslash S_{1,m}$.

Since $r(u_i|S_{1,m}) \neq r(u_j|S_{1,m})$, we have $r(u_i|S) \neq r(u_j|S)$. Similarly, since $r(v_i|S_{1,n}) \neq r(v_j|S_{1,n})$, we have $r(v_i|S) \neq r(v_j|S)$.

We next show that $r(u_i|S) \neq r(v_j|S)$ for each vertex $u_i \in V(P_m) \backslash S_{1,m}$ and for each vertex $v_j \in V(C_n) \backslash S_{1,n}$. Since $m, n \geq 7$, we have $\gamma, \alpha \geq 3$. If $\gamma = \alpha$, then we are done. Now, suppose $\gamma \neq \alpha$. Without loss of generality, let $\gamma = \alpha + t$ for some positive integer t. Observe that for any vertex $u_i \in V(P_m) \backslash S_{1,m}$ and any vertex $v_j \in V(C_n) \backslash S_{1,n}$, we have

$$r(u_i|S_{1,m}) = (d(u_i, s_1), \ldots, d(u_i, s_\alpha), d(u_i, s_{\alpha+1}), \ldots, d(u_i, s_{\alpha+t}))$$

and $r(v_j|S_{1,m}) = (\underbrace{1, \ldots, 1}_{\alpha \text{ times}}, \underbrace{1, \ldots, 1}_{t \text{ times}})$.

Since $\alpha \geq 3$, $d(u_i, s_k) \neq 1$ for some $1 \leq k \leq \alpha$. Hence, $r(u_i|S_{1,m}) \neq r(v_j|S_{1,m})$, which implies that $r(u_i|S) \neq r(v_j|S)$. It follows that the metric representations of the vertices of $V(P_m + C_n) \backslash S$ with respect to S are distinct. Hence, S is a resolving set for $P_m + C_n$ and it follows that $\beta(P_m + C_n) \leq \beta(F_{1,m}) + \beta(W_{1,n})$.

To prove that $\beta(P_m+C_n) \geq \beta(F_{1,m})+\beta(W_{1,n})$, it suffices to show that there is no resolving set for $P_m + C_n$ of cardinality less than or equal to $\gamma+\alpha-1$.

Suppose S' is a resolving set of $P_m + C_n$ of cardinality less than or equal to $\gamma+\alpha-1$. By Remark 3, we must have $S' = P'_m \cup C'_n$ where P'_m and C'_n are nonempty subsets of $V(P_m)$ and $V(C_n)$, respectively. If $1 \leq |P'_m| < \gamma$, then by Theorem 3, there exist $u_i, u_j \in V(P_m)\backslash P'_m$ such that $r(u_i|P'_m) = r(u_j|P'_m)$. Consequently, we have $r(u_i|S) = r(u_j|S)$, which is a contradiction. If $|P'_m| \geq \gamma$, then $|C'_n| \leq \alpha - 1$. By Theorem 2, there exist $v_i, v_j \in V(C_n)\backslash C'_n$ such that $r(v_i|C'_n) = r(v_j|C'_n)$. Consequently, we have $r(v_i|S) = r(v_j|S)$, which is also a contradiction. If $1 \leq |C'_n| < \alpha$, then by Theorem 2, there exist $v_i, v_j \in V(C_n)\backslash C'_n$ such that $r(v_i|C'_n) = r(v_j|C'_n)$. Consequently, we have $r(v_i|S) = r(v_j|S)$, which is a contradiction. If $|C'_n| \geq \alpha$, then $|P'_m| \leq \gamma - 1$. By Theorem 3 there exist $u_i, u_j \in V(P_m)\backslash P'_m$ such that $r(u_i|P'_m) = r(u_j|P'_m)$. Consequently, we have $r(u_i|S) = r(u_j|S)$, which is also a contradiction. Hence, there exists no resolving set of $P_m + C_n$ of cardinality less than or equal to $\gamma+\alpha-1$. Therefore, $\beta(P_m + C_n) \geq \beta(F_{1,m}) + \beta(W_{1,n})$ and so $\beta(P_m + C_n) = \beta(F_{1,m}) + \beta(W_{1,n})$. □

References

1. Buczkowski, P.S., Chartrand, G., Poisson, C., Zhang, P.: On k-dimensional graphs and bases. Period. Math. Hungar. **46**(1), 9–15 (2003)
2. Cáceres, J., Hernando, C., Mora, M., Pelayo, I.M., Puertas, L.M., Seara, C.: On the metric dimension of some families of graphs. Electron. Notes Disc. Math. **22**, 129–133 (2005)
3. Chartrand, G., Eroh, L., Johnson, M., Oellerman, O.: Resolvability in graphs and metric dimension of a graph. Discrete Appl. Math. **105**, 99–113 (2000)
4. Harary, F., Melter, R.A.: On the metric dimension of a graph. Ars Combin. **2**, 191–195 (1976)
5. Shanmukha, B., Sooryanarayana, B., Harinath, K.S.: Metric dimension of wheels. Far East J. Appl. Math. **8**(3), 217–229 (2002)
6. Slater, P.J.: Leaves of trees. Congr. Numer. **14**, 549–559 (1975)
7. Sooryanarayana, B.: On the metric dimension of a graph. Indian J. Pure Appl. Math. **29**(4), 413–415 (1998)

Barnette's Conjecture Through the Lens of the Mod_kP Complexity Classes

Robert D. Barish(✉) and Akira Suyama

Graduate School of Arts and Sciences, University of Tokyo, Meguro-ku Komaba 3-8-1, Tokyo 153-8902, Japan
rbarish@ims.u-tokyo.ac.jp, suyama@dna.c.u-tokyo.ac.jp

Abstract. In circa 2006, Feder & Subi established that Barnette's 1969 conjecture, which postulates that all cubic bipartite polyhedral graphs are Hamiltonian, is true if and only if the Hamiltonian cycle decision problem for this class of graphs is polynomial time solvable (assuming $\mathcal{P} \neq \mathcal{NP}$). Here, we bridge the truth of Barnette's conjecture with the hardness of a related set of decision problems belonging to the Mod_kP complexity classes (not known to contain $\mathcal{NP}$), where we are tasked with deciding if an integer k fails to evenly divide the Hamiltonian cycle count of a cubic bipartite polyhedral graph. In particular, we show that Barnette's conjecture is true if there exists a polynomial time procedure for this decision problem when k can be any arbitrarily selected prime number. However, to illustrate the barriers for utilizing this result to prove Barnette's conjecture, we also show that the aforementioned decision problem is Mod_kP-complete $\forall k \in (2\mathbb{N}_{>0} + 1)$, and more generally, that unless $\mathcal{NP} = \mathcal{RP}$, no polynomial time algorithm can exist if k is not a power of two.

1 Introduction

We examine Barnette's 1969 conjecture [2,13] on the Hamiltonicity of cubic bipartite polyhedral (i.e. 3-vertex-connected planar [27]) graphs from the perspective of the Mod_kP classes of decision problems [4,14], which for any $k \in \mathbb{N}_{>1}$ corresponds to the set of all languages L of decision problems solvable in polynomial time on a nondeterministic Turing machine, under the acceptance condition that the number of accepting pathways modulo k for any string $s \in L$ is non-zero. As Barnette's conjecture has now been open for more than 50 years, an important point of motivation for us is that the Mod_kP classes allow us to "decompose" the question as to whether a given cubic bipartite polyhedral graph is Hamiltonian—which by a proof of Feder & Subi [9] is $\mathcal{NP}$-complete, to decide if and only if Barnette's conjecture is false—into a number of "moderated" decision problems whose answers yield only partial information about the cardinality $|C|$ of the graph's set of Hamiltonian cycles. Determining, for

This work was supported by a Grant-in-Aid for JSPS Research Fellow (18F18117 to R. D. Barish) from the Japan Society for the Promotion of Science.

J. Akiyama et al. (Eds.): JCDCGGG 2018, LNCS 13034, pp. 59–73, 2021.
https://doi.org/10.1007/978-3-030-90048-9_6

instance, that $|C| \equiv 0 \pmod{k}$ for some integer k does not tell us that $|C| = 0$ or $|C| \neq 0$, however it arguably does leave us knowing something more than before.

We remark that there is precedence for Mod_kP decision problems being unexpectedly easy for certain values of k. One well-known example of this comes from a discovery due to Valiant [32] that counting satisfying assignments of planar read-twice monotone $3SAT$ formula (i.e. $3SAT$ formula where the bipartite clause-variable graph is planar, each variable occurs at most twice, and all literals are positive) is polynomial time tractable over the finite field $\mathbb{Z}_7$. In this particular case, and completely unexpectedly, one is able to preserve the number of satisfying assignments modulo 7 in a reduction to the problem of counting weighted perfect matchings on a planar graph (where the Fisher-Kasteleyn-Temperley (FKT) algorithm [17] can be employed).

Here, we first show that, unless $P = NP$, Barnette's conjecture is true if there exists a polynomial time procedure which takes any prime k and a cubic bipartite polyhedral graph as input, and returns $True$ if and only if k fails to evenly divide the Hamiltonian cycle count for the input graph (Theorem 1). Next, to illustrate the barriers for utilizing this result to prove Barnette's conjecture, we show that deciding if some $k \in \mathbb{N}_{>1}$ evenly divides the Hamiltonian cycle count for a cubic bipartite polyhedral graph is not polynomial time tractable under standard complexity theoretic assumptions if k is not a power of two (Theorem 2 & Theorem 3). Concerning the remaining cases, while a theorem of Kotzig tells us that all cubic bipartite graphs must have zero or an even number of Hamiltonian cycles [5], the complexity of the aforementioned decision problem is left open in all instances where $k = 2^n$ $(n \in \mathbb{N}_{>1})$.

2 Preliminaries

2.1 Graph Theoretic Preliminaries

Let G be a graph which is connected and simple (i.e. multi-edge and loop-free). We refer to G as being *cubic* if and only if its vertices are uniformly of degree 3, and refer to G as being *subcubic* if and only if the maximum degree of any vertex is at most 3. We call G *bipartite* if and only if its vertices are 2-colorable, or equivalently, if and only if it is free of odd length cycles. We call G *planar* if and only if it is embeddable in $\mathbb{R}^2$ (equivalently, $\mathbb{S}^2$) without edge crossings. We call G *k-vertex-connected* (respectively, *k-edge-connected*) if and only if for all pairs of vertices $\{v_a, v_b\}$, there exist at least k vertex disjoint paths (respectively, at least k edge disjoint paths) between v_a and v_b [22], and remark that a k-vertex-connected graph is often simply referred to as being *k-connected.* Here, if G is a cubic graph, it is k-vertex-connected if and only if it is k-edge-connected. If G is both 3-vertex-connected and planar, by Steinitz's theorem [27] we may refer to it as a *polyhedral* graph.

We refer to G as being *essentially-k-vertex-connected* (respectively, *essentially-k-edge-connected*) if the size of the minimum vertex cut (respectively, edge cut) necessary to decompose G into at least two connected components, where at least two such components contain an edge, is of size at least k. If G contains at least two vertex disjoint simple cycles, we refer to G as being *cyclically-k-vertex-connected* (respectively, *cyclically-k-edge-connected*) if the size of the minimum vertex cut (respectively, edge cut) necessary to decompose G into at least two connected components, where at least two such components contain a simple cycle, is of size at least k.

We make a few brief remarks concerning essential- and cyclic-k-(vertex or edge)-connectivity. To begin, we remark that any cubic graph $G \neq \{K_4, K_{3,3}\}$ will necessarily have at least two vertex disjoint simple cycles (see "Theorem 1.2" of [21]). If the minimum degree of a graph G is at least $\left(\frac{k}{2}+1\right)$, we have that G is essentially-k-edge-connected if and only if G is cyclically-k-edge-connected (see "Lemma 1" of [10]), implying that any cubic graph G is essentially-4-edge-connected if and only if it is cyclically-4-edge-connected (see "Corollary 1" of [10]). If $G \neq \{K_4, K_{3,3}, K_2 \square K_3\}$ is a cubic graph (where "$\square$" corresponds to the graph Cartesian product operation), G is cyclically-k-vertex-connected if and only if it is cyclically-k-edge-connected (see "Theorem 1.2" of [21], though note that the author neglects to mention the pathological case where $G = K_2 \square K_3$). Putting these latter two observations together, and further observing that for cubic $G \neq \{K_4, K_{3,3}, K_2 \square K_3\}$, G is essentially-4-edge-connected if and only if it is essentially-4-vertex-connected, we have that the notions of essential-4-vertex-connectivity, essential-4-edge-connectivity, cyclic-4-vertex-connectivity, and cyclic-4-edge-connectivity are equivalent on cubic graphs outside of three pathological cases (which correspond to all cubic graphs of order ≤ 6).

2.2 Complexity Theoretic Preliminaries

The complexity class $\mathcal{NP}$ (Nondeterministic Polynomial time) [8,16,18] consists of the set of all languages L of decision problems solvable in polynomial time on a nondeterministic Turing machine, or equivalently, the set of all languages L of decision problems where proofs are verifiable in polynomial time on a deterministic Turing machine. Completeness for the class $\mathcal{NP}$ is typically defined with respect to many-one (Karp) reductions [16].

The Mod_kP ($k \geq 2$) classes of decision problems [4,14] correspond to the set of all languages L of decision problems solvable in polynomial time on a nondeterministic Turing machine under the acceptance condition that the number of accepting pathways modulo k for any string $s \in L$ must be non-zero. While there is not, to our knowledge, a canonical reduction type used to define completeness for the class Mod_kP (for some integer k), in this work we will everywhere concern ourselves with completeness under many-one (Karp) reductions [16]. We remark that the complexity class Mod_2P is equivalent to the perhaps better known $\oplus P$ (Parity polynomial time) complexity class [12,23,29,30], first formally defined by Papadimitriou and Zachos [23], which consists of the set of all languages L of decision problems solvable in polynomial time on a nondeterministic Turing machine under the acceptance condition that there are an odd number of accepting pathways.

The complexity class $\mathcal{RP}$ (Randomized Polynomial time) consists of the set of all languages of decision problems solvable in polynomial time on a probabilistic Turing machine which reliably returns $False$ if no witness exists for the problem instance and otherwise returns $True$ with probability $\geq 1/2$. The complexity class $\mathcal{BPP}$ is defined similarly to $\mathcal{RP}$, with the exception that regardless of whether the answer is $False$ or $True$, the probability of the probabilistic Turing machine returning the correct answer is $\geq 2/3$. We remark that Impagliazzo [15] referred to the possibility that $\mathcal{NP} \subseteq \mathcal{BPP}$, which would imply $\mathcal{NP} = \mathcal{RP}$, as being a "moral equivalent" to the possibility that $\mathcal{P} = \mathcal{NP}$.

Concerning counting complexity, the class $\#P$ (pronounced "sharp" P) [29,30] is defined as the set of integer function problems of the form $f : \Sigma^* \longrightarrow \mathbb{N}$, where one is tasked with determining the number of accepting pathways for a nondeterministic Turing machine running in polynomial time on some input string x, and where the number of accepting pathways is at most exponential in the size of x. Completeness for the class $\#P$ can be established under either Turing reductions [29,30] or what are known as *many-one counting* ("weakly parsimonious") reductions [34] where, in reducing some integer counting problem f to another integer counting problem h, we have two polynomial time compatible functions $R_1 : \Sigma^* \longrightarrow \Sigma^*$ and $R_2 : \mathbb{N} \longrightarrow \mathbb{N}$, such that $f(x) = R_2(h(R_1(x)))$. Here, we call a many-one counting reduction *parsimonious* if R_2 is the identity function.

3 Theorems and Proofs

Theorem 1. If there exists a polynomial time procedure, which takes a prime k and a cubic bipartite polyhedral graph as input, and returns $True$ if and only if the Hamiltonian cycle count modulo k for the graph is non-zero, then Barnette's conjecture is true (unless $\mathcal{P} = \mathcal{NP}$).

Proof. Feder & Subi [9] proved that the Hamiltonian cycle decision problem on cubic bipartite polyhedral graphs is $\mathcal{NP}$-complete if and only if Barnette's 1969 conjecture [2,13] is false. It therefore suffices to show that, if there exists a polynomial time procedure $\mathcal{F}$, which takes an arbitrary specified prime k and returns $True$ if and only if the cardinality modulo k of the witness set W, for a string $r \in L$ of a language $L \in \mathcal{NP}$ is non-zero, then we have that $L \in P$.

To establish this latter statement, we begin by observing a lemma of Brightwell & Winkler [6] which says that the product of the set of primes lying between n and n^2 is at least $n! \times 2^n$ ($\forall n \geq 4$). This was originally used to show that the cardinality of a witness set, for an instance of SAT, can be uniquely recovered from the output of polynomially many modulo operations with primes. This lemma correspondingly implies that only polynomially many calls to the AKS primality test [1] for each integer $k = 1$ to n^2, and conditionally $\mathcal{F}$, suffices to prove that either $|W| = 0$ or $|W| > n! \times 2^n$. Here, as we can always determine an at most exponential upperbound $B_{|W|}$ for $|W|$—consider that the Cook-Levin theorem [8,16,18] gives a polynomial time parsimonious reduction from any $\mathcal{NP}$

language to an instance of SAT having at most $2^{(\#variables)}$ witnesses—we can simply test all primes on the interval $[1, n^2]$ where $n < \log_2(B_{|W|})$. Putting everything together, if any call to $\mathcal{F}$ for prime k is $True$, we have that $|W| \neq 0$, and if all calls to $\mathcal{F}$ for prime k return $False$, we have that $|W| = 0$. □

Theorem 2. Deciding if the number of Hamiltonian cycles on a cubic bipartite polyhedral graph modulo k is non-zero is Mod_kP-complete $\forall k \in (2\mathbb{N}_{>0} + 1)$.

Proof. We proceed by first showing a parsimonious reduction from $\#SAT$ to the problem of counting Hamiltonian cycles on a graph G, corresponding to a subdivision of a cubic bipartite polyhedral graph, where no two subdivided edges are adjacent. We will then identify all subdivided edges in G with a gadget Q, to create a graph H satisfying the following three requirements:

- (Requirement 1) H is a cubic bipartite polyhedral graph.
- (Requirement 2) Hamiltonian cycle traversals of H *induced* by Hamiltonian cycle traversals of G are multiplied by a power of two, which will have no effect on whether the Hamiltonian cycle count modulo $k \in (2\mathbb{N}_{>0} + 1)$ is non-zero.
- (Requirement 3) The number of Hamiltonian cycle traversals of H not *induced* by Hamiltonian cycle traversals of G has k as a factor, ensuring that the count for these "illegal" traversals is congruent to 0 modulo k.

To briefly clarify what it means when we say that a Hamiltonian cycle of H is *induced* by a Hamiltonian cycle of G, recall that we specified earlier that no two subdivided edges in G can be adjacent. Letting E_G and E_H correspond to the edge sets for the graphs G and H, respectively, observe that we can create a simple bijection between the set of edges $\Upsilon_G \subset E_G$ adjacent to two vertices of degree 3 in G, and the set of edges $\Upsilon_H \subset E_H$ disjoint from the gadgets Q in H, by assigning specific labels $l_1, l_2, \ldots \in L$ to each edge $e_i \in \Upsilon_G$, and then maintaining these labels during the identification operations to create H from G. Here, we say that a Hamiltonian cycle in H is *induced* by a Hamiltonian cycle in G if and only if both cycles have identical sets of labeled edges.

To now obtain a parsimonious reduction from $\#SAT$ to counting Hamiltonian cycles on a graph G, corresponding to a subdivision of a cubic bipartite polyhedral graph, where no two subdivided edges are adjacent, we will modify two gadgets from a circa 1976 reduction, due to Garey, Johnson, & Tarjan [11]. These were used to reduce arbitrary instances of $3SAT$ to the problem of deciding the existence of a Hamiltonian cycle on a cubic 3-vertex-connected planar graph. We remark that, due to the technical complexity of this construction, we will treat it largely as a "black box", and for further details refer the reader to the publication [11], where it was originally introduced. We also refer the reader to a more recent recent 2003 paper by Liśkiewicz, Ogihara, and Toda [19] that modifies the Garey, Johnson, & Tarjan reduction [11] to prove that counting Hamiltonian cycles on cubic 2-vertex-connected planar graphs is $\#P$-complete, under a special form of many-one ("weakly parsimonious") reductions.

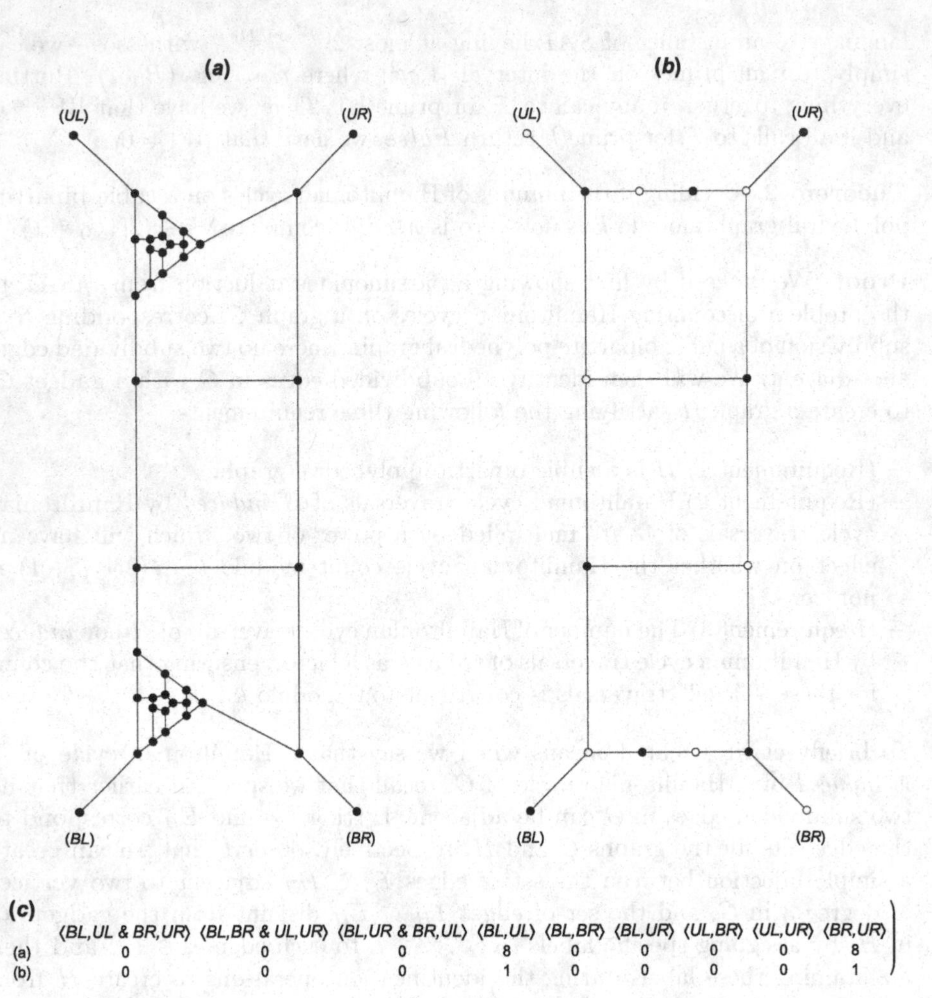

	⟨BL,UL & BR,UR⟩	⟨BL,BR & UL,UR⟩	⟨BL,UR & BR,UL⟩	⟨BL,UL⟩	⟨BL,BR⟩	⟨BL,UR⟩	⟨UL,BR⟩	⟨UL,UR⟩	⟨BR,UR⟩
(a)	0	0	0	8	0	0	0	0	8
(b)	0	0	0	1	0	0	0	0	1

Fig. 1. Exclusive-OR (XOR) gadget variants: **(a)** original Garey-Johnson-Tarjan XOR gadget [11] employed in their $\mathcal{NP}$-completeness proof for the problem of deciding the existence of a Hamiltonian cycle on cubic 3-vertex-connected planar graphs; **(b)** XOR gadget variant (this work) used to parsimoniously reduce $\#SAT$ to the problem of counting Hamiltonian cycles on subdivisions of cubic bipartite polyhedral graphs, where no two subdivided edges are adjacent; **(c)** counts for the number of ways in which a Hamiltonian cycle can traverse each XOR gadget via ingressing and egressing the indicated pairs of Upper-Left (UL), Upper-Right (UR), Bottom-Left (BL), and Bottom-Right (BR) vertices. Counts listed in (c) in the column designated, for example, "<BL,UL & BR,UR>", correspond to instances where a Hamiltonian cycle flows through the gadget twice, once by ingressing from vertex (BL) and egressing from vertex (UL) (or vice versa), and once by ingressing at vertex (BR) and egressing at vertex (UR) (or vice versa). Counts in the column designed, for example, "<BL,UR>", correspond to instances where a Hamiltonian cycle flows through the gadget only once, ingressing at vertex (BL) and egressing at vertex (UR) (or vice versa).

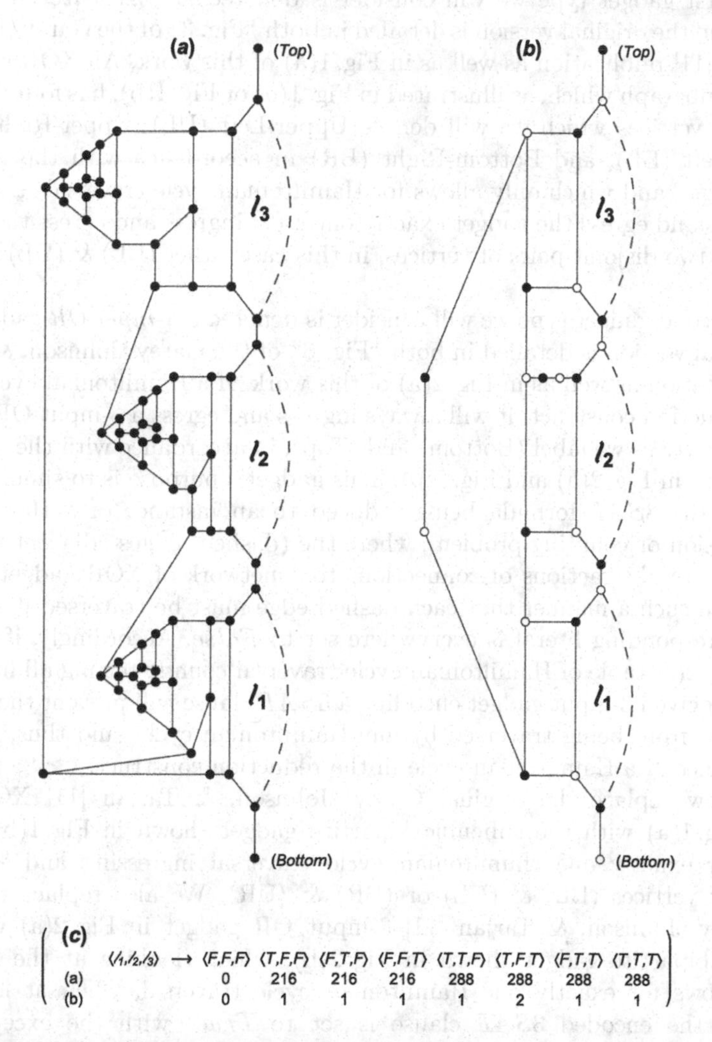

Fig. 2. Variants of the 3-input OR gadget: **(a)** original Garey-Johnson-Tarjan 3-input OR gadget [11] employed in their $\mathcal{NP}$-completeness proof for the problem of deciding the existence of a Hamiltonian cycle on cubic 3-vertex-connected planar graphs; **(b)** 3-input OR gadget variant (this work) used to parsimoniously reduce $\#SAT$ to the problem of counting Hamiltonian cycles on subdivisions of cubic bipartite polyhedral graphs, where no two subdivided edges are adjacent; **(c)** counts for the number of ways in which a Hamiltonian cycle can traverse each 3-input OR gadget via ingressing and egressing the indicated (Bottom) and (Top) vertices under all possible $True$ and $False$ assignments for literals $\{l_1, l_2, l_3\}$. Here, literals $\{l_1, l_2, l_3\}$ are simulated as being set to $False$ (respectively, $True$) if the dashed loop to the immediate right of the label for a given literal must be traversed (respectively, cannot be traversed) by all Hamiltonian cycles.

The first gadget type we will consider is denoted an *Exclusive-OR* (XOR) gadget, and the original version is detailed in both "Fig. 3" of the Garey, Johnson, & Tarjan [11] publication as well as in Fig. 1(a) of this work. An XOR gadget is simply a subgraph which, as illustrated in Fig. 1(a) or Fig. 1(b), has four outgoing edges—to vertices which we will denote Upper-Left (UL), Upper-Right (UR), Bottom-Left (BL), and Bottom-Right (BR) in accordance with the provided embeddings—and which only allows for Hamiltonian cycle traversals that both: (1) ingress and egress the gadget exactly once; (2) ingress and egress the gadget via one of two disjoint pairs of vertices, in this case either (BL) & (UL) or (BR) & (UR).

The second gadget type we will consider is denoted a 3-*input OR* gadget, and the original version is detailed in both "Fig. 6" of the Garey, Johnson, & Tarjan [11] publication as well as in Fig. 2(a) of this work. If a Hamiltonian cycle exists in the reduction construct, it will always ingress and egress a 3-input OR gadget from the vertices we label (Bottom) and (Top), in accordance with the provided embeddings in Fig. 2(a) and Fig. 2(b). This gadget's purpose is to simulate each clause in the $3SAT$ formula being reduced to an instance of a Hamiltonian cycle decision or counting problem, where the (dashed) edges adjacent to labels $\{l_1, l_2, l_3\}$ are abstractions of connections to a network of XOR gadgets, wired together in such a manner that each dashed edge must be traversed if and only if the corresponding literal is everywhere set to $False$. Accordingly, if we look at the Fig. 2(c) table of Hamiltonian cycle traversal counts, setting all literals to $False$ in a given 3-input gadget encoding a $3SAT$ clause will prevent the 3-input OR gadget from being traversed by any Hamiltonian cycle, and thus, prohibit the existence of a Hamiltonian cycle in the reduction construct.

We now replace the original Garey, Johnson, & Tarjan [11] XOR gadget in Fig. 1(a) with the subcubic bipartite gadget shown in Fig. 1(b), which allows for exactly one Hamiltonian cycle traversal ingressing and egressing via either vertices (BL) & (UL) or (BR) & (UR). We also replace the original Garey, Johnson, & Tarjan [11] 3-input OR gadget in Fig. 2(a) with the subcubic bipartite gadget shown in Fig. 2(b), which (looking at the Fig. 2(c) table) allows for exactly one Hamiltonian cycle traversal, when at least one literal in the encoded $3SAT$ clause is set to $True$—with the exception of a single pathological case where $\{l_1, l_2, l_3\} = \{True, False, True\}$. Here, we avoid this pathological case by first parsimoniously reducing $3SAT$ to a variant of satisfiability denoted $3SAT^*$, where we employ two dummy variables to replace all clauses of the form $\{x, y, z\}$ with a logically equivalent expression $(x \vee \neg u \vee \neg w) \wedge (y \vee z \vee u) \wedge (\neg y \vee \neg u \vee \neg w) \wedge (\neg z \vee \neg u \vee \neg w) \wedge (w \vee w \vee w)$, having the property that the truth assignment $\{l_1, l_2, l_3\} = \{True, False, True\}$ will never occur in any satisfying assignment. Similar methods of avoiding certain literal $False$ and $True$ assignment permutations in clauses are elaborated upon by Liśkiewicz, Ogihara, and Toda [19] and Valiant [31].

Following the details of the Garey, Johnson, & Tarjan [11] reduction in either their original publication or in Liśkiewicz, Ogihara, and Toda [19], we can observe that the Hamiltonian cycle traversal properties of our Fig. 1(b) XOR and Fig. 2(b) 3-input OR gadgets will yield a parsimonious reduction transitively from $\#SAT$ to the problem of counting Hamiltonian cycles on subcubic 2-vertex-connected planar graphs where no two subdivided edges are adjacent. With a little more work we can also observe—as Plesník [25] originally had in extending the Garey, Johnson, & Tarjan [11] reduction to the bipartite case—that the fact both gadgets are bipartite implies that the final reduction construct will be bipartite. Thus, we now have our parsimonious reduction from $\#SAT$ to the problem of counting Hamiltonian cycles on a subdivision of a cubic bipartite polyhedral graph, where no two subdivided edges are adjacent.

For the last step of the proof argument, we can observe that all subdivided edges in our reduction construct from the previous step are of the form shown in Fig. 3(a), having the Hamiltonian cycle traversal properties shown in the Fig. 3(d) table. Here, taking two copies of the Fig. 3(b) ladder graph with $n \in (2\mathbb{N}_{>0} + 1)$ rungs, where we say that a ladder graph has n rungs if it is isomorphic to the grid graph $G_{2,n}$, we can join these graphs to create the Fig. 3(c) gadget Q. Observe now that identifying all Fig. 3(a) subgraphs in the reduction construct (corresponding to subdivided edges) with the gadget Q in the manner shown in Fig. 3(c)—respecting the adjacent Upper-Left (UL), Upper-Right (UR), Bottom-Left (BL), and Bottom-Right (BR) vertices—will ensure that the resulting graph is a cubic bipartite polyhedral graph, satisfying (Requirement 1). Observe also that the Fig. 3(c) gadget multiplies the number of Hamiltonian cycle traversals possible for the Fig. 3(a) subgraph by the prime factor 2, thus satisfying (Requirement 2).

Finally, observe that the Fig. 3(c) gadget will multiply the number of Hamiltonian cycles traversing the gadget in a manner disallowed in the Fig. 3(a) subgraph by either $2(n-1)$ or $(n-1)^2$, and thus, $\forall k \in (2\mathbb{N}_{>0} + 1)$ we can specify $n \in 2\mathbb{N}_{>0}+1$ such that $(n-1) = k$. This will ensure that the set of Hamiltonian cycles which traverse the Fig. 3(c) gadget in a manner disallowed on the Fig. 3(a) subgraph will be congruent to 0 modulo k, and therefore, that their contribution to the cardinality $|C|$ of the set of Hamiltonian cycles will not effect the value of $|C|$ modulo k, satisfying (Requirement 3).

Putting everything together, we have that the hardness of deciding if the cardinality of the Hamiltonian cycle set for a cubic bipartite polyhedral graph modulo k is non-zero is equivalent to the hardness of deciding if the number of satisfying assignments for an arbitrary $3SAT$ (or transitively, SAT) formula modulo k is non-zero, yielding the theorem. □

Theorem 3. *No polynomial time algorithm exists for deciding if the number of Hamiltonian cycles on a cubic bipartite polyhedral graph modulo $k \in \mathbb{N}_{>1}$ is non-zero if k is not a power of two (unless $\mathcal{NP} = \mathcal{RP}$).*

Proof. We proceed by modifying the Theorem 2 proof argument to allow for k to be an even integer. The existing Theorem 2 proof argument fails in this case as a consequence of the fact that the final step of identifying subdivided edges with the Fig. 3(c) gadget will necessarily multiply the cardinality of any set of Hamiltonian cycles by some power of two. Here, if k is an even integer,

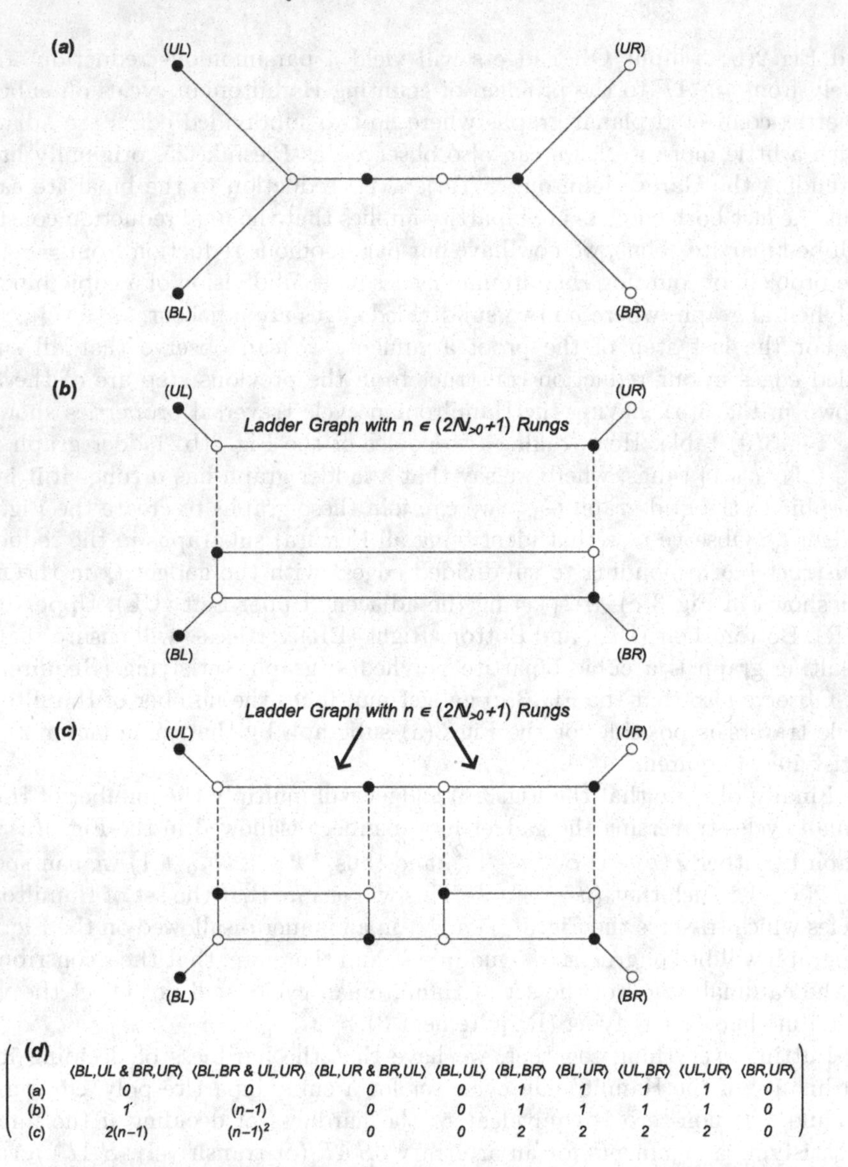

(d)	⟨BL,UL & BR,UR⟩	⟨BL,BR & UL,UR⟩	⟨BL,UR & BR,UL⟩	⟨BL,UL⟩	⟨BL,BR⟩	⟨BL,UR⟩	⟨UL,BR⟩	⟨UL,UR⟩	⟨BR,UR⟩
(a)	0	0	0	0	1	1	1	1	0
(b)	1	$(n-1)$	0	0	1	1	1	1	0
(c)	$2(n-1)$	$(n-1)^2$	0	0	2	2	2	2	0

Fig. 3. Illustration and Hamiltonian cycle traversal properties of subdivided edges in cubic graphs and the gadgets they are identified with in the Theorem 2 proof argument: **(a)** an example of a subdivided edge in a subcubic bipartite graph; **(b)** a cubic bipartite ladder graph with $n \in (2\mathbb{N}_{>0} + 1)$ rungs; **(c)** the gadget Q composed of two ladder graphs having $n \in (2\mathbb{N}_{>0} + 1)$ rungs, which can be identified in the manner shown—respecting the adjacent Upper-Left (UL), Upper-Right (UR), Bottom-Left (BL), and Bottom-Right (BR) vertices—with the subdivided edges of a subcubic bipartite graph (where no two subdivided edges are adjacent) to make the graph cubic bipartite; **(d)** counts for the number of ways in which a Hamiltonian cycle can traverse each subgraph or gadget via ingressing and egressing the indicated pairs of (UL), (UR), (BL), and (BR) vertices. Here, column labels in (d) have the same meaning as column labels in Fig. 1(c).

it necessarily has 2 as a prime factor. Therefore, we can no longer guarantee that the number of Hamiltonian cycles in the final reduction construct modulo k is non-zero if and only if this holds true for the number of satisfying instances of the encoded $\#SAT$ formula. Consider the example where we set $k = 8$ and encode a $\#SAT$ formula with exactly 12 satisfying assignments, then observe that $12 \equiv 4 \pmod 8$ while $12 \times 2 \equiv 0 \pmod 8$. Furthermore, noting Tutte's proof [28]—a novel proof of a result attributed to C. A. B. Smith [28]—that all edges of a cubic graph must lie on zero or an even number of Hamiltonian cycles, we can see that this is in fact a "fundamental" property of any cubic graph gadget attempting to mimic the Hamiltonian cycle traversal properties of a subdivided edge of a cubic graph (see Fig. 3(a) and the Hamiltonian cycle traversals it allows in the Fig. 3(d) table).

To circumvent the aforementioned issue we need to ensure that k has some prime factor disjoint from the set of prime factors for the cardinality $|C|$ of Hamiltonian cycles in the final reduction construct. We do this by initially proceeding via a reduction from a variant of satisfiability known as Unambiguous-SAT [33], where we are promised the existence of at most one satisfying assignment for a specified Boolean formula. Otherwise following the Theorem 2 proof argument, we can reduce arbitrary instances of Unambiguous-SAT to the problem of counting Hamiltonian cycles on subdivisions of cubic bipartite polyhedral graphs where no two subdivided edges are adjacent, and where, by virtue of the fact that the reduction is parsimonious, we are promised the existence of at most one Hamiltonian cycle.

When we now identify the subdivided edges in this reduction construct with the Fig. 3(c) gadget to obtain a cubic bipartite polyhedral graph, we can ensure that 2 is the only prime factor for the cardinality $|C|$ of Hamiltonian cycles. Accordingly, if $k \in \mathbb{N}_{>1}$ is not a power of two, we have that deciding if $|C|$ modulo k is non-zero is equivalent in hardness to the task of counting satisfying assingments for an arbitrary instance of an Unambiguous-SAT formula modulo k. Finally, observe that this latter problem is equivalent to deciding if an arbitrary Unambiguous-SAT formula has either zero or exactly one satisfying assignment, and recall that by the Valiant-Vazirani theorem [33], no polynomial time algorithm can exist for Unambiguous-SAT unless $\mathcal{NP} = \mathcal{RP}$.

Putting everything together, if $k \in \mathbb{N}_{>1}$ is not a power of two, and unless $\mathcal{NP} = \mathcal{RP}$, there can be no polynomial time procedure for deciding if the number of Hamiltonian cycles modulo k on a cubic bipartite polyhedral graph is non-zero. □

4 Note Concerning Essentially-4-(Vertex,Edge)-Connected or Cyclically-4-(Vertex,Edge)-Connected Cubic Bipartite Polyhedral Graphs

We remark that there exists an inductive definition, due to Batagelj [3], for the class of cubic bipartite polyhedral graphs as the set of graphs which can be constructed by any finite number of applications of the following two surgeries or "generating rules" with the 3-cube as the initial object or "basis": (surgery "s") identification of an arbitrary vertex as the BW_3 subgraph (see surgery "s" in "Fig. 10" of [3]), resulting in the addition of 6 vertices and 9 edges; (surgery "q") a C_4 cycle addition surgery (see surgery "q" in "Fig. 10" of [3]), resulting in the addition of 4 vertices and 4 edges. Here, we can observe that the 3-cube is cyclically-4-vertex-connected, and that the surgery q will not perturb this property. Therefore, the set of graphs which can be constructed by any finite number of applications of the surgery "q" with the 3-cube as the "basis" will generate a subset of all cyclically-4-vertex-connected cubic bipartite polyhedral graphs, which, according to our remarks in Sect. 2.1, must also be cyclically-4-edge-connected, essentially-4-vertex-connected, and essentially-4-edge-connected.

We argue that it may be worth asking if Barnette's 1969 conjecture [2,13] holds in the case where we restrict our attention to the aforementioned set of cyclically-4-vertex-connected cubic bipartite polyhedral graphs. Towards addressing this question, we ask the reader to observe that the Garey, Johnson, & Tarjan [11] reduction construct, and our variation of this reduction construct given in the Theorem 2 proof argument, will be cyclically-4-vertex-connected if and only if this property holds for the component XOR and OR gadgets. Here, while the original Garey, Johnson, & Tarjan [11] XOR and OR gadgets (illustrated in Fig. 1(a) and Fig. 2(a), respectively) do not have this property, our variations of these gadgets (illustrated in Fig. 1(b) and Fig. 2(b), respectively) were designed to be cyclically-4-vertex-connected after everywhere identifying subdivided edges with the Fig. 3(c) gadget. This yields the following lemma:

Lemma 1. The proof arguments given in Theorem 2 and Theorem 3 hold in the case where it is required that all cubic bipartite polyhedral graphs are cyclically-4-vertex-connected (equivalently, cyclically-4-edge-connected, essentially-4-vertex-connected, or essentially-4-edge-connected).

5 Note Concerning the Parsimonious Reduction Given in the Theorem 2 Proof Argument

While our Theorem 2 proof argument required a novel parsimonious reduction transitively from $\#SAT$ to the problem of counting Hamiltonian cycles on subdivisions of cubic bipartite polyhedral graphs, we remark that there exist other earlier examples of parsimonious reductions from $\#SAT$ to counting Hamiltonian cycles on subcubic planar graphs. The very first such reduction we are aware of appears in a circa 2002 undergraduate senior thesis of Takahiro Seta [26], which we were able to independently reconstruct and verify. Seemingly unaware of Seta's work, in circa 2005 Valiant [31] also provides a reduction from $\#SAT$ to counting Hamiltonian cycles on subcubic 2-vertex-connected planar graphs to establish that counting Hamiltonian cycles modulo two on this class of graphs is $\oplus P$-complete, and furthermore, to give an example of a problem complete for the class $\oplus P$ under many-one (Karp) reductions.

Valiant's reduction [31] is arguably much simpler than Seta's [26], however, we were not able to verify the stated properties of one of the gadgets. Specifically, looking at the partial illustration of a 3-input OR gadget in "Fig. 2" of [31], we do not understand why a Hamiltonian cycle cannot traverse the gadget via ingressing and egressing only the A_3 and B_2 labeled edges, which would allow the gadget to be traversed even if all encoded literals are set to $False$. It is true that this issue vanishes under the requirement (stated in the "Figure 2" caption of [31]) that: *"...the global construction is such that any Hamiltonian circuit entering at some A_i must next traverse B_i before it can traverse an A_j or B_j with $j \neq i$"*. However, it is also clearly stated in Valiant's "Theorem 3" that the author follows *"...the proof for the regular degree three case given by Liskiewicz, Ogihara and Toda, [LOT03] [ref.* [19] *in this manuscript], which itself elaborates on [GJT76] [ref.* [11] *in this manuscript] ...*", and we are unaware of any property of either construction which would enforce the aforementioned constraint.

6 Note Concerning Attempts at Proving Barnette's Conjecture

While there is currently no universally accepted proof of Barnette's conjecture (see e.g. [24]), we are aware of at least one attempt at a proof due to Cahit [7]. Specifically, in "Theorem 4" of the aforementioned work, Cahit claims that his "Carve-Cubic-Planar" algorithm will always generate a Hamiltonian cycle for a cubic bipartite polyhedral graph. We make no statement here as to the correctness of this claim, however, we note that "Theorem 5" of the same work states: *"For every non-hamiltonian 3-connected cubic planar graph, Algorithm 1 terminates with a cycle of length $n-1$."*, where n corresponds to the order of the input graph. This latter claim contradicts a theorem of Lu [20] that the circumference of a cubic 3-vertex-connected planar graph G, denoted $c(G)$, satisfies the inequality $c(G) \leq n - (n + \frac{49}{4})^{\frac{1}{2}} + (\frac{5}{2})$ for infinitely many integers n. Here, as $n - \left(n - (n + \frac{49}{4})^{\frac{1}{2}} + (\frac{5}{2})\right) \geq 2$ for $n \geq 8$, both theorems cannot be correct.

7 Concluding Remarks

We have shown how to use the Mod_kP classes of decision problems to reformulate Barnette's 1969 conjecture [2,13], namely that every cubic bipartite polyhedral graph is Hamiltonian, as a question of the computational tractability of a set of decision problems where we ask whether the Hamiltonian cycle count for a given cubic bipartite polyhedral graph is evenly divisible by some integer k. As this opens the door to some of these decision problems being polynomial time tractable—notably, for $k = 2$ a theorem of Kotzig tells us that all cubic bipartite graphs must have zero or an even number of Hamiltonian cycles [5]—we have limited this possibility to instances where $k = 2^n$ for $n \geq 2$ (assuming

$\mathcal{NP} \neq \mathcal{RP}$). We leave the reader with the conjecture that, regardless of the truth of Barnette's conjecture, there exist polynomial time algorithms in each of these remaining cases.

References

1. Agrawal, M., Kayal, N., Saxena, N.: PRIMES is in P. Ann. Math. **160**(2), 781–793 (2004)
2. Barnette, D.: Conjecture 5. In: Tutte, W.T. (ed.) Recent Problems in Combinatorics, p. 343. Academic Press, New York (1969)
3. Batagelj, V.: Inductive definition of two restricted classes of triangulations. Discrete Math. **52**(2–3), 113–121 (1984)
4. Beigel, R., Gill, J., Hertramp, U.: Counting classes: thresholds, parity, mods, and fewness. In: Choffrut, C., Lengauer, T. (eds.) STACS 1990. LNCS, vol. 415, pp. 49–57. Springer, Heidelberg (1990). https://doi.org/10.1007/3-540-52282-4_31
5. Bosák, J.: Hamiltonian lines in cubic graphs. In: Fiedler, M. (ed.) Proceedings of the International Seminar on Graph Theory and Applications, Rome, July 1966; Appearing in Theory of Graphs, pp. 35–46. Gordon & Breach, New York (1967)
6. Brightwell, G., Winkler, P.: Counting linear extensions. Order **8**(3), 225–242 (1991)
7. Cahit, I.: Algorithmic proof of Barnette's conjecture. arXiv:0904.3431, pp. 1–13 (2009)
8. Cook, S.A.: The complexity of theorem-proving procedures. In: Proceedings of the 3rd Annual ACM Symposium on Theory of Computing (STOC), pp. 151–158 (1971)
9. Feder, T., Subi, C.: On Barnette's conjecture. Rep. TR06-015, Electronic Colloquium on Computational Complexity (ECCC) (2006)
10. Fleischner, H., Jackson, B.: A note concerning some conjectures on cyclically 4-edge connected 3-regular graphs. Ann. Discrete Math. **41**, 171–178 (1989)
11. Garey, M.R., Johnson, D.S., Tarjan, R.E.: The planar Hamiltonian circuit problem is NP-complete. SIAM J. Comput. **5**(4), 704–714 (1976)
12. Goldschlager, L.M., Parberry, I.: On the construction of parallel computers from various bases of boolean functions. Theoret. Comput. Sci. **43**(1), 43–58 (1986)
13. Grunbaum, B.: Polytopes, graphs, and complexes. Bull. Amer. Math. Soc. **76**, 1131–1201 (1970)
14. Hertrampf, U.: Relations among Mod-classes. Theoret. Comput. Sci. **74**(3), 325–328 (1990)
15. Impagliazzo, R.: A personal view of average-case complexity. In: Proceedings of the 10th Annual Structure in Complexity Theory Conference (SCT), pp. 134–147 (1995)
16. Karp, R.M.: Reducibility among combinatorial problems. In: Miller, R.E., Thatcher, J.W. (eds.)Complexity of Computer Computations, pp. 85–103. Plenum Press, New York (1972)
17. Kasteleyn, P.W.: Graph theory and crystal physics. In: Harary, F. (ed.) Graph Theory and Theoretical Physics, pp. 43–110. Academic Press, New York (1967)
18. Levin, L.A.: Universal search problems. Probl. Peredachi Inf. **9**(3), 265–266 (1973)
19. Liśkiewicz, M., Ogihara, M., Toda, S.: The complexity of counting self-avoiding walks in subgraphs of two-dimensional grids and hypercubes. Theoret. Comput. Sci. **304**(1–3), 129–156 (2003)

20. Lu, X.: A note on 3-connected cubic planar graphs. Discrete Math. **310**(13–14), 2054–2058 (2010)
21. McCuaig, W.: Edge reductions in cyclically k-connected cubic graphs. J. Combin. Theory Ser. B **56**(1), 16–44 (1992)
22. Menger, K.: Zur allgemeinen kurventheorie. Fundam. Math. **10**(1), 96–115 (1927)
23. Papadimitriou, C.H., Zachos, S.K.: Two remarks on the power of counting. In: Cremers, A.B., Kriegel, H.-P. (eds.) GI-TCS 1983. LNCS, vol. 145, pp. 269–275. Springer, Heidelberg (1982). https://doi.org/10.1007/BFb0036487
24. Pirzada, S., Shah, M.A.: Construction of Barnette graphs whose large subgraphs are non-Hamiltonian. Acta Univ. Sapientiae Math. **11**(2), 363–370 (2019)
25. Plesník, J.: The NP-completeness of the Hamiltonian cycle problem in bipartite cubic planar graphs. Acta Math. Univ. Comenian. **42–43**, 271–273 (1983)
26. Seta, T.: The complexities of puzzles, Cross Sum and their Another Solution Problems (ASP). Senior Thesis, Department of Infomation Science, the Faculty of Science, the University of Tokyo (2002)
27. Steinitz, E.: Polyeder und raumeinteilungen. Encyklopadie der mathematischen Wissenschaften. Bd. III-1B, Hft. 9, pp. 1–139 (1922)
28. Tutte, W.T.: On Hamiltonian circuits. J. London Math. Soc. **s1–21**(2), 98–101 (1946)
29. Valiant, L.G.: The complexity of computing the permanent. Theoret. Comput. Sci. **8**(2), 189–201 (1979)
30. Valiant, L.G.: The complexity of enumeration and reliability problems. SIAM J. Comput. **8**(3), 410–421 (1979)
31. Valiant, L.G.: Completeness for parity problems. In: Wang, L. (ed.) COCOON 2005. LNCS, vol. 3595, pp. 1–8. Springer, Heidelberg (2005). https://doi.org/10.1007/11533719_1
32. Valiant, L.G.: Accidental algorithms. In: Proceedings of the 47th Annual Symposium on Foundations of Computer Science (FOCS), pp. 509–517 (2006)
33. Valiant, L.G., Vazirani, V.V.: NP is as easy as detecting unique solutions. Theoret. Comput. Sci. **47**(1), 85–93 (1986)
34. Zankó, V.: #P-completeness via many-one reductions. Int. J. Found. Comput. Sci. **2**(1), 77–82 (1991)

Hamiltonicity of Graphs on Surfaces in Terms of Toughness and Scattering Number – A Survey

Kenta Ozeki(✉)

Faculty of Environment and Information Sciences, Yokohama National University, Yokohama, Japan
ozeki-kenta-xr@ynu.ac.jp

Abstract. This paper aims to survey Hamiltonicity of graphs on surfaces, including stronger (e.g. Hamiltonian-connectedness) and weaker (e.g. containing Hamiltonian paths and spanning trees with certain conditions) properties. Toughness and scattering number conditions are necessary conditions for graphs to have such properties. Since every k-connected graph on a surface F^2 satisfies some toughness and scattering number condition, we can expect that "every k-connected graph on a surface F^2 satisfies the property $\mathcal{P}$". We explain which triple $(k, F^2, \mathcal{P})$ makes the statement true from the viewpoint of toughness and scattering number of graphs.

Keywords: Hamiltonian cycles · Spanning trees · Toughness · Scattering number · Plane graphs

MSC 2010: 05C05 · 05C10 · 05C38 · 05C45

1 Introduction

This paper aims to survey Hamiltonicity of graphs on surfaces, including stronger (e.g. Hamiltonian-connectedness) and weaker (e.g. containing Hamiltonian paths and spanning trees with certain conditions) properties. From the viewpoint of toughness and scattering number of graphs, we discuss which triple $(k, F^2, \mathcal{P})$ makes the statement "every k-connected graph on a surface F^2 satisfies the property $\mathcal{P}$" true.

2 Preliminaries

2.1 Hamiltonicity

A cycle or a path C in a graph G is said to be *Hamiltonian* if C visits all vertices in G. A graph with a Hamiltonian cycle is said to be *Hamiltonian*. Hamiltonicity of graphs is one of the important topics in graph theory because of its relation

J. Akiyama et al. (Eds.): JCDCGGG 2018, LNCS 13034, pp. 74–95, 2021.
https://doi.org/10.1007/978-3-030-90048-9_7

to several problems such as the Travelling Salesman Problem. In particular, Hamiltonicity of planar graphs or graphs on surfaces is related to the Four Color Theorem, and hence this has attracted many researchers. For example, Tait [97] claimed to show in 1880 that "if every 3-connected planar cubic graph is Hamiltonian, then every planar graph can be colored by four colors". Note that it was later shown that the assumption of Tait's theorem does not hold (see [105] for example), and hence this approach does not arrive at a solution to the Four Color Problem. However, since then, Hamiltonicity of graphs on surfaces has attracted wide attention.

For example, Whitney [107] proved in 1931 that every 4-connected plane triangulation is Hamiltonian, and Tutte [103] extended this result to 4-connected planar graphs;

Theorem 1 (Tutte [103]**).** *Every 4-connected planar graph is Hamiltonian.*

Many results have been shown, and we will introduce those in this survey.

A graph G is said to be *Hamiltonian-connected* if for every pair of distinct vertices x and y in G, there is a Hamiltonian path in G between x and y. Note that any Hamiltonian-connected graph is Hamiltonian and any Hamiltonian graph contains a Hamiltonian path. We denote properties related to Hamiltonicity by $\mathcal{P}_{\mathrm{H}}, \mathcal{P}_{\mathrm{HP}}, \mathcal{P}_{\mathrm{HC}}$ as follows;

$$\begin{aligned} \mathcal{P}_{\mathrm{H}} &: \text{Contains a Hamiltonian cycle,} \\ \mathcal{P}_{\mathrm{HP}} &: \text{Contains a Hamiltonian path,} \\ \mathcal{P}_{\mathrm{HC}} &: \text{Is Hamiltonian-connected.} \end{aligned}$$

2.2 The Condition (∗), Toughness and Scattering Number

We denote by $c(H)$ the number of components of a graph H. For a graph G and real numbers a and b, consider the following condition;

$$(*) \qquad c(G - S) \leq a|S| + b \text{ for any } S \subseteq V(G) \text{ with } c(G - S) \geq 2.$$

The *toughness* of a graph G is the minimum value of a satisfying (∗) with $b = 0$. Chvátal [26] (see also [27]) in 1973 introduced the concept of toughness and conjectured that there exists a constant t_0 such that every graph satisfying (∗) with $(a, b) = (\frac{1}{t_0}, 0)$ is Hamiltonian. This conjecture is still open. Chvátal first conjectured $t_0 > 3/2$ suffices, but some counterexamples were found (e.g. see [38]). Bauer, Broersma and Veldman [8] showed that there exist non-Hamiltonian graphs satisfying (∗) with $(a, b) = (\frac{4}{9}, \frac{11}{9})$. Thus, $t_0 \geq 9/4$ even if Chvátal's conjecture holds. See [7,15] for more information on toughness.

On the other hand, the *scattering number* of a graph G is defined as the minimum of b satisfying (∗) with $a = 1$. This concept was defined by Jung [57], as the "additive dual" for the concept of toughness.

As pointed out in [26,57], it is easy to see the following necessary condition for Hamiltonicity.

Proposition 2. *Let G be a graph.*

- *If G contains a Hamiltonian path, then G satisfies $(*)$ with $(a, b) = (1, 1)$.*
- *If G is Hamiltonian, then G satisfies $(*)$ with $(a, b) = (1, 0)$.*
- *If G is Hamiltonian-connected, then G satisfies $(*)$ with $(a, b) = (1, -1)$.*

In this survey, we focus on several properties which have the condition $(*)$ as a necessary condition, such as those in Proposition 2. For a property $\mathcal{P}$, we define the real numbers $a_{\mathcal{P}}$ and $b_{\mathcal{P}}$ as follows;

$$(a_{\mathcal{P}}, b_{\mathcal{P}}) = \inf \Big\{ (a, b) : \text{Every graph with the property } \mathcal{P} \text{ satisfies } (*) \text{ with } (a, b) \Big\},$$

where the infimum is taken in the lexicographical order. Note that some properties $\mathcal{P}$ do not have $a_{\mathcal{P}}$ and $b_{\mathcal{P}}$, but we do not consider such properties $\mathcal{P}$ in this survey.

For example, Proposition 2 implies $(a_{\mathcal{P}_{\mathrm{H}}}, b_{\mathcal{P}_{\mathrm{H}}}) \leq (1, 0)$, where $\leq$ represents the lexicographical order. On the other hand, consider the complete bipartite graph $K_{m,m}$ with $m \geq 2$, which is Hamiltonian. We see that for any $a < 1$ and any real number b, if we take a large integer m satisfying $(1 - a)m > b$, then $c(K_{m,m} - S) = m > am + b = a|S| + b$, where S is the set of vertices contained in one partite set. Furthermore, for any $b < 0$, $c(K_{m,m} - S) = m > m + b = |S| + b$ for the same vertex set S. Thus, for any $a < 1$ and any b, and for $a = 1$ and any $b < 0$, there exists a Hamiltonian graph that does not satisfy $(*)$ with (a, b). These show that $(a_{\mathcal{P}_{\mathrm{H}}}, b_{\mathcal{P}_{\mathrm{H}}}) = (1, 0)$. Similarly, we see $(a_{\mathcal{P}_{\mathrm{HP}}}, b_{\mathcal{P}_{\mathrm{HP}}}) = (1, 1)$ and $(a_{\mathcal{P}_{\mathrm{HC}}}, b_{\mathcal{P}_{\mathrm{HC}}}) = (1, -1)$, by using $K_{m,m+1}$, and the graph obtained from the complete bipartite graph $K_{m,m-1}$ by adding all possible edges in the larger bipartite set for a sufficiently large integer m, respectively.

In addition, the property of containing a spanning tree with bounded maximum degree is also related to the condition $(*)$. For $k \geq 2$, a *k-tree* is a tree with maximum degree at most k. A Hamiltonian path is nothing but a spanning 2-tree, and hence a spanning k-tree is related to Hamiltonicity. We denote by $\mathcal{P}_{k\text{-tree}}$ the property of containing a spanning k-tree. For this property, it is known that $(a_{\mathcal{P}_{k\text{-tree}}}, b_{\mathcal{P}_{k\text{-tree}}}) = (k - 1, 1)$, see [80, Proposition 5] for a proof. We refer the readers to a survey [80] on spanning trees. Note that Win [108] showed that every graph satisfying $(*)$ with $(a, b) = (k - 2, 2)$ contains a spanning k-tree, and a short proof is found in [37].

As an extension of a spanning k-tree, we define the following concept. For $k \geq 2$ and a tree T, the *total excess of T from k*, denoted by $\mathrm{te}(T, k)$, is defined as

$$\mathrm{te}(T, k) = \sum_{v \in V(T)} \max\{\deg_T(v) - k, 0\},$$

where $\deg_T(v)$ is the degree of a vertex v in T. Note that a tree T satisfies $\mathrm{te}(T, k) = 0$ if and only if T is a k-tree. Furthermore, a tree T satisfies $\mathrm{te}(T, 2) \leq t$ if and only if T has at most $(t+2)$ leaves. We denote by $\mathcal{P}_{(k,t)\text{-tree}}$ the property of containing a spanning tree for which the total excess from k is at most t. Similar to the case of $\mathcal{P}_{k\text{-tree}}$, it can be shown that $(a_{\mathcal{P}_{(k,t)\text{-tree}}}, b_{\mathcal{P}_{(k,t)\text{-tree}}}) = (k - 1, t + 1)$.

A *spanning closed k-walk*, which is a spanning closed walk visiting every vertex at most k times, is related to a spanning k-tree. By the definition, a spanning closed 1-walk is nothing but a Hamiltonian cycle. Note that if a graph contains a spanning k-tree, then it contains a spanning closed k-walk, and if a graph contains a spanning closed k-walk, then it contains a spanning $(k+1)$-tree (see [55]). Let $\mathcal{P}_{k\text{-walk}}$ be the property of containing a spanning closed k-walk. Then we see that $(a_{\mathcal{P}_{k\text{-walk}}}, b_{\mathcal{P}_{k\text{-walk}}}) = (k, 0)$. Recall that $(a_{\mathcal{P}_{k\text{-tree}}}, b_{\mathcal{P}_{k\text{-tree}}}) = (k-1, 1)$. Thus, we expect that the property of containing a spanning closed k-walk is close to the property of containing a spanning $(k+1)$-tree, while the former is stronger than the latter. This expectation is discussed in Sect. 5.2 (see also [55]).

2.3 Graphs on Surfaces

We refer to Mohar and Thomassen [67] for the detail of surfaces and graphs on them.

A *surface* F^2 is a connected compact 2-dimensional manifold without boundary. By the classification of surfaces, F^2 is either an orientable surface of genus $t \geq 0$, denoted by S_t, or a nonorientable surface of genus $k \geq 1$, denoted by N_k. The *Euler genus* of a surface F^2, denoted by $g(F^2)$, is defined as

$$g(F^2) = \begin{cases} 2t & \text{if } F^2 \text{ is the orientable surface } S_t, \\ k & \text{if } F^2 \text{ is the nonorientable surface } N_k. \end{cases}$$

Note that S_0 is the sphere, N_1 is the projective plane, S_1 is the torus, and N_2 is the Klein bottle, where their Euler genera are $0, 1, 2$ and 2, respectively.

A graph G *on a surface* F^2 means an embedding of G in F^2, that is, a drawing of G on F^2 in which the edges do not cross each other. We can regard plane graphs also as graphs on the sphere through the inverse of the stereographic projection. Euler's formula states any graph G on F^2 satisfies

$$|V(G)| - |E(G)| + |F(G)| \geq 2 - g(F^2).$$

(Note that if G is 2-cell embedding in F^2, that is, every face is homeomorphic to a disk, then the equality holds). The following is am important relation between the condition $(*)$ and graphs on surfaces.

Proposition 3 (Schmeichel and Bloom [92]). *Let $k \geq 3$ and let G be a k-connected graph on a surface F^2. Then G satisfies $(*)$ with $(a, b) = \left(\dfrac{2}{k-2}, \dfrac{2g(F^2)-4}{k-2}\right)$.*

Proof. We follow the proof by Goddard, Plummer and Swart [47].

Let $S \subseteq V(G)$, and let c be the number of components in $G-S$ (that is, $c = c(G-S)$). We construct the graph H by contracting each component in $G-S$ to one vertex, deleting all obtained multiple edges and loops, and deleting all edges connecting two vertices in S. Then H is a bipartite graph still on the surface F^2. Hence, Euler's formula directly implies $|E(H)| \leq 2(|S|+c)+2g(F^2)-4$. Since G is k-connected, each component in $G-S$ has at least k neighbors in S, and hence $|E(H)| \geq kc$. These two inequalities imply that $c(G-S) = c \leq \frac{2}{k-2}|S|+\frac{2g(F^2)-4}{k-2}$, as desired. □

Consider a triangulation H of F^2, and take the *face subdivision* of H, that is, add a new vertex to each face and connect it to all the vertices on the facial cycle. The obtained graph, call it G, is a triangulation of F^2. By Euler's formula, we can calculate $c(G - V(H)) = |F(H)| = 2|V(H)| + 2g(F^2) - 4$. This means that G satisfies the bound obtained in Proposition 3 with equality when $S = V(H)$ for $k = 3$. Similarly, if we begin with quadrangulations with minimum degree at least 2 as H, then we can construct triangulations G of F^2 satisfying the bound obtained in Proposition 3 with equality when $S = V(H)$ for $k = 4$. To construct 5-connected ones, we choose graphs H more carefully. Some pentangulations with minimum degree at least 3 work. The following shows that Proposition 3 is in some sense best possible.

Proposition 4. *For $k = 3, 4$ or 5 and any surface F^2, there exist infinitely many k-connected triangulations G on F^2 with $S \subseteq V(G)$ such that $c(G-S) \geq 2$ and*

$$c(G - S) = \frac{2}{k-2}|S| + \frac{2g(F^2) - 4}{k-2}.$$

To show the same conclusion for the case $k = 6$, we need to begin with some hexangulations of a surface F^2 with minimum degree at least 3. (Otherwise, after the face subdivision, the vertices in H will have degree at most 4, and hence it is not 5-connected.) Such hexangulations exist for any surfaces F^2 with $g(F^2) \geq 2$, but do not exist for the sphere and the projective plane. Similarly, for the case $k = 7$, we need to begin with a heptangulation with minimum degree at least 4. Note that for each surface F^2, there are only finitely many heptangulations. In fact, there are only finitely many 7-connected graphs on F^2. Thus, we do not consider the case $k \geq 7$ in this survey.

On the other hand, it is worthwhile to notice that for $k = 3, 4$ or 5 and any surface F^2, there exist infinitely many k-connected triangulations G on F^2 that satisfy the bound obtained in Proposition 3 with strict inequality for any $S \subset V(G)$. For example, consider the triangulation obtained from the face subdivision of a 4-connected triangulation by removing some added vertices. We can check that such triangulations satisfy the above condition. We leave the details to the readers.

A closed curve γ on a surface is said to be *contractible* if γ bounds a 2-cell region on F^2; Otherwise, γ is *essential.* The *representativity*, rep(G), of a graph G on a surface F^2 is the minimum number of intersecting points of G and γ, where γ ranges over all essential closed curves on F^2. (If F^2 is the sphere, then we define rep(G) $= \infty$.) Graphs on surfaces with sufficiently large representativity are sometimes informally called *locally planar* graphs, and are known to have properties similar to plane graphs. We obtain the next proposition in the same way as Proposition 3.

Proposition 5. *Let $k \geq 3$, $a_0 > \frac{2}{k-2}$ and b_0 be a real number. Then for any surface F^2, there exists a constant r such that every k-connected graph G on F^2 with rep$(G) \geq r$ satisfies $(*)$ with $(a, b) = (a_0, b_0)$.*

Note that for each k, Archdeacon, Hartsfield and Little [1] constructed a k-connected triangulation G_k, embedded in some orientable surface F^2 with $S \subseteq V(G_k)$ such that $\text{rep}(G_k) \geq k$ and $c(G_k - S) \geq k|S|$. Note that the genus of F^2 depends on k, and this shows that the constant r in Proposition 5 must depend on the surface F^2.

2.4 Purpose of the Study

For $k \geq 3$, this survey investigates a surface F^2, and several properties $\mathcal{P}$, such as $\mathcal{P}_{\text{H}}$, $\mathcal{P}_{\text{HP}}$, $\mathcal{P}_{\text{HC}}$, whether the statement

every k-connected graph on F^2 satisfies $\mathcal{P}$

holds or not. In fact, this statement does not hold for some triples $(k, F^2, \mathcal{P})$, since as in Proposition 2, some properties $\mathcal{P}$ requires $(*)$ with $(a, b) = (a_\mathcal{P}, b_\mathcal{P})$ as a necessary condition, but Proposition 4 shows that for some surface F^2, there are k-connected graphs on F^2 that do not satisfy $(*)$ with $a_\mathcal{P}$ and $b_\mathcal{P}$.

We expect, however, that the statement above holds for $(k, F^2, \mathcal{P})$ if all k-connected graphs on F^2 satisfy $(*)$ with $(a, b) = (a_\mathcal{P}, b_\mathcal{P})$. We introduce several results and conjectures, which support this expectation. In this sense, the relation between Hamiltonicity and the connectivity of graphs on surfaces seems to be based on the condition $(*)$. See also the survey [32].

For example, consider the sphere S^2, which satisfies $g(S^2) = 0$. Proposition 4 implies that there are infinitely many 4-connected graphs on the sphere S^2 satisfying the equality in $(*)$ with $(a, b) = (1, -2)$. Since $b_{\mathcal{P}_\text{H}} = 0 \geq -2 = g(S^2) - 2$, all 4-connected graphs on the sphere S^2 satisfy $(*)$ with $(a, b) = (a_{\mathcal{P}_\text{H}}, b_{\mathcal{P}_\text{H}})$, which is a necessary condition to be $\mathcal{P}_\text{H}$. Thus, one can expect that all 4-connected graphs on the sphere S^2 are Hamiltonian. In fact, Tutte showed that this expectation indeed holds, see Theorem 1.

This situation can be seen for several triples $(k, F^2, \mathcal{P})$. We discuss which triples make the statement "every k-connected graph on F^2 satisfy $(*)$ with $(a, b) = (a_\mathcal{P}, b_\mathcal{P})$ satisfies $\mathcal{P}$" true.

2.5 Remark on Non-Hamiltonian Planar Graphs

As in the previous subsection, our aim is to show some properties, related to Hamiltonicity, in k-connected graphs on a surface, in terms of the condition $(*)$. One may wonder if instead of the k-connected condition, we can directly use the condition $(*)$ and guarantee Hamiltonicity; however, this does not seem to behave properly.

It is well-known that every graph satisfying $(*)$ with $(a, b) = (t, 0)$ is $\lceil 2/t \rceil$-connected (see [26]). Thus, Theorem 1 directly implies that every planar graph, satisfying $(*)$ with $(a,b) = (t,0)$ for $t < \frac{2}{3}$, is Hamiltonian. Nishizeki [72] constructed non-Hamiltonian plane triangulations satisfying $(*)$ with $(a,b) = (1,0)$. This was later extended to non-Hamiltonian planar graphs satisfying $(*)$ with $(a,b) = (\frac{2}{3},0)$ by Harant [51], and also to non-Hamiltonian plane triangulations satisfying $(*)$ with $(a,b) = (\frac{2}{3},\frac{2}{3})$ by Owens [75]. (Whether there exists a non-Hamiltonian plane triangulation satisfying $(*)$ with $(a,b) = (\frac{2}{3},0)$ is an open problem.) In this sense, Theorem 1 on 4-connected planar graphs is the essential result, rather than a statement in terms of condition $(*)$.

3 4-Connectedness and Properties $\mathcal{P}$ with $a_{\mathcal{P}} = 1$

In this section, we focus on 4-connected graphs on surfaces. By Proposition 3,

> every 4-connected graph on a surface F^2 satisfies $(*)$ with $(a,b) = (1, g(F^2) - 2)$,

and by Proposition 4, there are infinitely many 4-connected graphs on F^2 that satisfy the equality in $(*)$ with $(a,b) = (1, g(F^2) - 2)$. Therefore, for properties $\mathcal{P}$ with $a_{\mathcal{P}} = 1$, such as $\mathcal{P}_{\mathrm{H}}$, $\mathcal{P}_{\mathrm{HP}}$ and $\mathcal{P}_{\mathrm{HC}}$, there are some surfaces F^2 such that every 4-connected graph on F^2 satisfies $(*)$ with $(a,b) = (a_{\mathcal{P}}, b_{\mathcal{P}})$. On the other hand, depending on the value of $b_{\mathcal{P}}$, Proposition 4 implies that there are some surfaces F^2 such that some 4-connected graphs on F^2 do not satisfy $(*)$ with $(a,b) = (a_{\mathcal{P}}, b_{\mathcal{P}})$.

For example, consider the torus T^2, which satisfies $g(T^2) = 2$. Proposition 4 implies that there are infinitely many 4-connected graphs on the torus T^2 satisfying the equality in $(*)$ with $(a,b) = (1,0)$. Since $b_{\mathcal{P}_{\mathrm{HC}}} = -1 < 0 = g(T^2) - 2$, those are not Hamiltonian-connected. For this reason, we will deal with only the triple $(4, F^2, \mathcal{P})$ that satisfy $a_{\mathcal{P}} = 1$ and $b_{\mathcal{P}} \geq g(F^2) - 2$.

3.1 Hamiltonian Paths, Hamiltonian Cycles and Hamiltonian-Connectedness

We first focus on the properties $\mathcal{P}_{\mathrm{H}}$, $\mathcal{P}_{\mathrm{HP}}$ and $\mathcal{P}_{\mathrm{HC}}$. We summarize the known results and conjectures in Table 1. As explained above, depending on the Euler genus $g(F^2)$, some surfaces F^2 may admit 4-connected graphs on F^2 that do not satisfy $(*)$ with $(a,b) = (a_{\mathcal{P}}, b_{\mathcal{P}})$. When such graphs exist, we use "×" for the corresponding cell in Table 1, which means that there exist infinitely many counterexamples to the corresponding statement. The cell with "○" means that the corresponding statement holds, as shown in the reference for the cell. (e.g. Thomassen [101] proved that every 4-connected planar graph is Hamiltonian-connected.) The cell with "?" means that the statement remains a conjecture.

Table 1. Hamiltonicity of 4-connected graphs on a surface.

	$\mathcal{P}_{\mathrm{HP}}$ $(b_{\mathcal{P}_{\mathrm{HP}}} = 1)$	$\mathcal{P}_{\mathrm{H}}$ $(b_{\mathcal{P}_{\mathrm{H}}} = 0)$	$\mathcal{P}_{\mathrm{HC}}$ $(b_{\mathcal{P}_{\mathrm{HC}}} = -1)$
Sphere(plane) $g(F^2) = 0$	○	○ Tutte [103]	○ Thomassen [101]
Projective plane $g(F^2) = 1$	○	○ Thomas & Yu [98][a]	○ Kawarabayashi & Ozeki [62][b]
Torus $g(F^2) = 2$	○ Thomas, Yu & Zang [100]	? Grünbaum [50] Nash-Williams [71]	×
Klein bottle $g(F^2) = 2$	?	?	×
N_3 $g(F^2) = 3$	?	×	×
Other surfaces $g(F^2) \geq 4$	×	×	×

[a] This statement was known as Grünbaum's conjecture [50] before Thomas and Yu [98].
[b] This statement was known as Dean's conjecture [29] before Kawarabayashi and Ozeki [62]

As expliained in Sect. 2.1, Whitney [107] proved that every 4-connected plane triangulation is Hamiltonian before Tutte's theorem for 4-connected planar graphs (Theorem 1). Note that later, Tutte [104] himself gave a simpler proof.

Chiba and Nishizeki [23] pointed out an omission in Thomassen's proof [101], stating that every 4-connected planar graph is Hamiltonian-connected, and fixed it. See also [73, Chapter 10]. Other proofs for the theorem can be found in [65, 76]. Based on Thomassen's proof, Chiba and Nishizeki [24] presented an algorithm to find a Hamiltonian cycle in 4-connected planar graphs, which was an improvement of the algorithms in [2, 49]. More recent algorithms can be found in [10, 93, 94].

As in Table 1, there are several results on Hamiltonicity of 4-connected graphs on surfaces, but still some open problems remain. The following is the most important conjecture in this field.

Conjecture 6 (Grünbaum [50], **Nash-Williams** [71]**).** *Every 4-connected graph on the torus is Hamiltonian.*

The difficulty of Conjecture 6 seems to come from the proof method for 4-connected graphs on the sphere and the projective plane to be $\mathcal{P}_{\mathrm{H}}$. All results have been shown by proving a stronger property $\mathcal{P}'$ than $\mathcal{P}_{\mathrm{H}}$ (e.g. the property of containing a Hamiltonian cycle passing through a prescribed edge), which satisfies $(a_{\mathcal{P}'}, b_{\mathcal{P}'}) = (1, -1)$, because of inductive arguments. This method does not work for the toroidal case, however, since there are infinitely many 4-connected graphs on the torus that do not satisfy $(*)$ with $(a, b) = (1, -1)$ as in Proposition 4. Thus, we cannot expect to use the same idea.

Note that Proposition 3 implies that for a 4-connected graph G on the torus, G does not satisfy $(*)$ with $(a, b) = (1, -1)$, if and only if the equality in $(*)$ with $(a, b) = (1, 0)$ holds for some $S \subseteq V(G)$. Considering this situation, Fujisawa, Nakamoto and Ozeki proved the following theorem.

Theorem 7 (Fujisawa, Nakamoto and Ozeki [41,69]**).** *Let G be a 4-connected graph on the torus. If the equality in $(*)$ with $(a,b) = (1,0)$ holds for some $S \subseteq V(G)$, then G is Hamiltonian.*

Thus, to remedy Conjecture 6, it suffices to focus on 4-connected graphs on the torus that satisfy $(*)$ with $(a,b) = (1,-1)$. We may be able to use the method of proof that has been succeeded for the case of the sphere and the projective plane. Conjecture 6 is still open, but the author expects that it will be resolved using this argument.

3.2 Planar Graphs and Properties $\mathcal{P}$ with $(a_{\mathcal{P}}, b_{\mathcal{P}}) = (1,-2)$

If we restrict ourselves to 4-connected planar graphs G, then Proposition 3 implies that G satisfies $(*)$ with $(a,b) = (1,-2)$. Since $(a_{\mathcal{P}_{\mathrm{HC}}}, b_{\mathcal{P}_{\mathrm{HC}}}) = (1,-1)$, we expect that every 4-connected planar graph has properties stronger than being Hamiltonian-connected, such that they require $(*)$ with $(a,b) = (1,-2)$ as a necessary condition. In fact, the following results have been obtained:

- Every 4-connected planar graph G with at least 5 vertices satisfies for any $S \subseteq V(G)$ with $|S| \leq 2$, $G - S$ is Hamiltonian[1]. (This statement was conjectured by Plummer [83], and shown by Thomas and Yu [98]).
- For any two vertices x, y and an edge e with $e \neq xy$ in a 4-connected planar graph G, there exists a Hamiltonian path in G connecting x and y through e. (This statement was shown by Sanders [89]).
- For any vertex x in a 4-connected planar graph G, $G - x$ is Hamiltonian-connected. (This statement can be shown using a result by Sanders [89]).
- For vertices u_1, u_2, v_1, v_2, possibly $u_i = v_j$ for some $i, j \in \{1,2\}$, $G + \{u_1u_2, v_1v_2\}$ contains a Hamiltonian cycle through both u_1u_2 and v_1v_2. (This statement was shown by Ozeki and Vràna [79].)

If we denote these properties as $\mathcal{P}_{2\mathrm{H}}$, $\mathcal{P}_{\mathrm{eHC}}$, $\mathcal{P}_{1\mathrm{HC}}$ and $\mathcal{P}_{2\mathrm{eHC'}}$, respectively, then we obtain $(a_{\mathcal{P}_{2\mathrm{H}}}, b_{\mathcal{P}_{2\mathrm{H}}}) = (a_{\mathcal{P}_{\mathrm{eHC}}}, b_{\mathcal{P}_{\mathrm{eHC}}}) = (a_{\mathcal{P}_{1\mathrm{HC}}}, b_{\mathcal{P}_{1\mathrm{HC}}}) = (a_{\mathcal{P}_{2\mathrm{eHC'}}}, b_{\mathcal{P}_{2\mathrm{eHC'}}}) = (1,-2)$ in the same way as in Proposition 2, and the graph obtained from the complete bipartite graph $K_{m,m+2}$ by adding all possible edges in the larger bipartite set, for a sufficiently large integer m.

Note that any graph with the property $\mathcal{P}_{2\mathrm{eHC'}}$ satisfies the properties $\mathcal{P}_{\mathrm{eHC}}$ and $\mathcal{P}_{1\mathrm{HC}}$ (the case when $u_1 = x$, $u_2 = y$ and $v_1v_2 = e$ corresponds to $\mathcal{P}_{\mathrm{eHC}}$, and the case when $u_2 = v_1 = x$ corresponds to $\mathcal{P}_{1\mathrm{HC}}$); however, it is not known whether any graph with the property $\mathcal{P}_{2\mathrm{eHC'}}$ satisfies the property $\mathcal{P}_{2\mathrm{H}}$. At least there exist infinitely many graphs that satisfy the property $\mathcal{P}_{2\mathrm{H}}$, but which do not satisfy the property $\mathcal{P}_{2\mathrm{eHC'}}$.

A *figure-eight* graph consists of two edge-disjoint cycles sharing exactly one vertex. The shared vertex is called its *center*. Rosenfeld [86] proved that every

[1] This statement is related to Malkevitch's conjecture [66] stating that every 4-connected planar graph G contains a cycle of length l for any $3 \leq l \leq |V(G)|$, possibly except for $l = 4$. See also [21].

4-connected planar graph contains a figure-eight graph as a spanning subgraph with the prescribed vertex as a center. This property $\mathcal{P}_8$ is also related to the condition (*), which satisfies $(a_{\mathcal{P}_8}, b_{\mathcal{P}_8}) = (1, 0)$. Thus, we may expect properties stronger than $\mathcal{P}_8$ for 4-connected planar graphs.

3.3 Surfaces F^2 with $g(F^2) \geq 4$ and Properties $\mathcal{P}$ with $b_{\mathcal{P}} > 0$

As in Table 1, for any surface F^2 with $g(F^2) \geq 4$, there exist 4-connected graphs on F^2 without Hamiltonian paths. This is shown by Proposition 4 and the fact $b_{\mathcal{P}_{\mathrm{HP}}} = 1 < g(F^2) - 2$. However, we would like to find some properties that may be weaker than the property $\mathcal{P}_{\mathrm{HP}}$ but are related to Hamiltonicity. In this subsection, we focus on $\mathcal{P}_{(2,t)\text{-tree}}$, which is equivalent to the property of containing a spanning tree with at most $(t+2)$ leaves. Recall that $(a_{\mathcal{P}_{(2,t)\text{-tree}}}, b_{\mathcal{P}_{(2,t)\text{-tree}}}) = (1, t+1)$. Since every 4-connected graph G on F^2 satisfies (*) with $(a, b) = (1, g(F^2) - 2)$ (by Proposition 3), we expect the following conjecture, which was posed by the author, to hold.

Conjecture 8 (Ozeki [77]). *For every 4-connected graph G on a surface F^2 with $g(F^2) \geq 3$, there exists a spanning tree with at most $(g(F^2) - 1)$ leaves.*

Recently, using an idea similar to the one used in the proof of Theorem 7, Nakamoto and Ozeki [70] showed a partial solution for 4-connected graphs on a surface F^2 with $g(F^2) \geq 3$ such that the equality in (*) with $(a, b) = (1, g(F^2) - 2)$ holds for some $S \subseteq V(G)$.

Considering Proposition 5, every 4-connected locally planar graph on a surface satisfies (*) with $a = 2$. Ellingham, Gao [33] and Yu [106] focused on this fact and proved that every 4-connected locally planar graph on a surface contains a spanning 3-tree and a spanning closed 2-walk, respectively. Mohar (see [67, P. 181]) posed the problem whether every 4-connected locally planar graph on a surface contains a spanning tree with few leaves, which corresponds to Conjecture 8.

Problem 9 (Mohar, see [67]). *For any surface F^2 with $g(F^2) \geq 3$, does there exist a constant $r = r(F^2)$ such that every 4-connected graph G on F^2 with $rep(G) \geq r$ contains a spanning 3-tree with $O(g(F^2))$ leaves?*

Böhme, Mohar and Thomassen [12] showed a statement weaker than Problem 9 stating that for any surface F^2 with $g(F^2) \geq 3$ and any $\varepsilon > 0$, there exists a constant $r = r(F^2, \varepsilon)$ such that every 4-connected graph G on F^2 with $\mathrm{rep}(G) \geq r$ contains a spanning 3-tree with at most $\varepsilon|V(G)|$ leaves.

4 5-Connectedness and Properties $\mathcal{P}$ with $a_{\mathcal{P}} \geq \frac{2}{3}$

In this section, we focus on 5-connected graphs on surfaces. By Proposition 3,

any 5-connected graph on F^2 satisfies (*) with $(a, b) = \left(\frac{2}{3}, \frac{2g(F^2)-4}{3}\right)$.

Regarding Hamiltonicity of 4-connected graphs on surfaces, there are some open problems represented by "?" in Table 1 and false statements represented by "×". However, assuming the stronger assumption 5-connectedness, we may obtain a result that guarantees Hamiltonicity. For example, Thomas and Yu [99] proved that Conjecture 6 holds if we replace "4-connected" in the assumption with "5-connected". See Table 2 for more results.

Note that before those results in Table 2, Barnette [5] and Brunet and Richter [18] proved that every 5-connected triangulation of the torus contains a Hamiltonian path, and a Hamiltonian cycle, respectively. Brunet, Nakamoto and Negami [17] proved the same conclusion holds for 5-connected triangulation of the Klein bottle.

We would like to guarantee Hamiltonicity of 5-connected graphs on a surface F^2 with $g(F^2) \geq 4$. As explained in the previous section, this is impossible with the assumption of "4-connectedness" by Proposition 4, but Proposition 5 implies that every 5-connected locally planar graph on a surface satisfies $(*)$ with $(a, b) = (1, 0)$, which is a necessary condition for $\mathcal{P}_{\mathrm{H}}$. Based on this fact, we expect that the following conjecture holds. (See [67, P. 182] and [106].)

Conjecture 10. *For any surface F^2 with $g(F^2) \geq 3$, there exists a constant $r = r(F^2)$ such that every 5-connected graph G on F^2 with $rep(G) \geq r$ is Hamiltonian.*

Conjecture 10 is still open. Yu [106] showed that Conjecture 10 is true if we restrict ourselves to triangulations. This was conjectured by Thomassen [102]. Kawarabayashi [59] showed that Yu's result holds even if we replace the conclusion by "Hamiltonian-connected".

Table 2. Hamiltonicity of 5-connected graphs on a surface.

	$\mathcal{P}_{\mathrm{HP}}$	$\mathcal{P}_{\mathrm{H}}$	$\mathcal{P}_{\mathrm{HC}}$
Sphere(plane) $g(F^2) = 0$	O	O	O
Projective plane $g(F^2) = 1$	O	O	O
Torus $g(F^2) = 2$	O	O Thomas & Yu [99]	O Kawarabayashi & Ozeki [63]
Klein bottle $g(F^2) = 2$	?	?	?
Others (Locally planar) $g(F^2) \geq 3$	?	? Thomassen [102]	?

On the other hand, Proposition 3 guarantees that every 5-connected graph on a surface satisfies $(*)$ with $(a, b) = (\frac{2}{3}, t)$ for some t. This seems to suggest that such graphs satisfy some properties $\mathcal{P}$ with $1 > a_{\mathcal{P}} \geq \frac{2}{3}$. As far as the author knows, there are no results on such properties, and hence we leave this as an open problem.

Problem 11. *Which properties $\mathcal{P}$ satisfy $1 > a_{\mathcal{P}} \geq \frac{2}{3}$? Does every 5-connected planar graph (or 5-connected graph on a surface) satisfy the property $\mathcal{P}$?*

5 3-Connectedness and Properties $\mathcal{P}$ with $a_{\mathcal{P}} = 2$

As explained in the previous sections, 4-connected graphs on surfaces have some properties related to Hamiltonicity. In this section, we consider the similarity for 3-connected graphs. By Proposition 3,

any 3-connected graph on F^2 satisfies $(*)$ with $(a, b) = (2, 2g(F^2) - 4)$,

and by Proposition 4, there are infinitely many 3-connected graphs on F^2 that satisfy the equality in $(*)$ with $(a, b) = (2, 2g(F^2) - 4)$. Therefore, it is natural to show properties $\mathcal{P}$ with $a_{\mathcal{P}} \leq 2$ in 3-connected graphs on surfaces.

5.1 Spanning Trees

We first focus on the property $\mathcal{P}_{(3,t)\text{-tree}}$, which satisfies $a_{\mathcal{P}_{(3,t)\text{-tree}}} = 2$, together with some works on $\mathcal{P}_{k\text{-tree}}$ and $\mathcal{P}_{(2,t)\text{-tree}}$. We summarize those results in Table 3. The cell with "◯" means that the corresponding statement holds with best possible value on k or t, while "△" means that the corresponding statement holds but the sharpness is still unknown.

Barnette [6] proved that every 3-connected planar graph contains a spanning 3-tree. Similar to the extension from $\mathcal{P}_{\mathrm{H}}$ to $\mathcal{P}_{\mathrm{HC}}$, this result is extended to show the property that for any vertex x, there exists a spanning 3-tree in which x is a leaf [4,68]. If we denote this property by $\mathcal{P}_{3\text{-tree}^+}$, then we see that $(a_{\mathcal{P}_{3\text{-tree}^+}}, b_{\mathcal{P}_{3\text{-tree}^+}}) = (2, -1)$. Since every 3-connected planar graph satisfies $(*)$ with $(a, b) = (2, -4)$ by Proposition 3, we may expect that a further stronger statement holds.

Proposition 5 implies that there exist infinitely many 3-connected triangulations on a surface F^2, that satisfy the equality in $(*)$ with $(a, b) = (2, 2g(F^2) - 4)$ for some $S \subseteq V(G)$. Since $(a_{\mathcal{P}_{3\text{-tree}}}, b_{\mathcal{P}_{3\text{-tree}}}) = (2, 1)$, we cannot expect that every 3-connected graph on a surface F^2 with $g(F^2) \geq 3$ contains a spanning 3-tree. Thus, we consider the existence of a spanning tree with higher (but bounded) maximum degree in such graphs. Yu [106] proved that for any surface F^2 with $g(F^2) \geq 3$, there exists a constant $r = r(F^2)$ such that every 3-connected graph G on F^2 with $\mathrm{rep}(G) \geq r$ contains a spanning 4-tree. Note that before Yu [106], Thomassen [102] proved the same statement for triangulations of orientable surfaces. Sanders and Zhao [91], and Ota and Ozeki [74] proved that every 3-connected graph G on F^2 with $g(F^2) \geq 3$ contains a spanning $\lceil \frac{2g(F^2)+4}{3} \rceil$-tree. (Sanders and Zhao [91] proved the case $g(F^2) \geq 38$, and later Ota and Ozeki [74] extended this to the case $g(F^2) \geq 3$.) Note that the upper bound "$\lceil \frac{2g(F^2)+4}{3} \rceil$" of the maximum degree in the obtained spanning tree is best possible. This is

Table 3. Spanning trees and closed walks in 3-connected graphs on a surface.

	$\mathcal{P}_{k\text{-tree}}$	$\mathcal{P}_{(2,t)\text{-tree}}$	$\mathcal{P}_{k\text{-walk}}$
Sphere(plane) $g(F^2)=0$	○ Barnette [6] $(k=3)$	△ N. O. O. [68] $\left(t=\frac{\lvert G\rvert-7}{3}\right)$	○ Gao & Richter [43] $(k=2)$
Projective plane $g(F^2)=1$	○ Barnette [4] $(k=3)$	○ N. O. O. [68] $\left(t=\frac{\lvert G\rvert-7}{3}\right)$	○ Gao & Richter [43] $(k=2)$
Torus/K-bottle $g(F^2)=2$	○ Barnette [4] $(k=3)$	○ N. O. O. [68] $\left(t=\frac{\lvert G\rvert-3}{3}\right)$	○ B. E. G. M. R. [16] $(k=2)$
Others (Locally planar) $g(F^2)\geq 3$	○ Yu [106] $(k=4)$	△ $\left(t=\frac{\lvert G\rvert+O(1)}{3}\right)$	○ Yu [106] $(k=3)$
Others $g(F^2)\geq 3$	○ Sanders & Zhao [91] Ota & Ozeki [74] $\left(k=\lceil\frac{2g(F^2)+4}{3}\rceil\right)$	?	○ Sanders & Zhao [91] Hasanvand [52] $\left(k=\lceil\frac{2g(F^2)+2}{3}\rceil\right)$

also obtained by condition $(*)$. Before [74], Ellingham [32] obtained a weaker conclusion.

Furthermore, similar to the extension mentioned in Sect. 3.3, those results on a spanning 3-tree were extended to a spanning tree with bounded total excess from 3 for the case of surfaces F^2 with $g(F^2) \geq 3$. Recall that $(a_{\mathcal{P}_{(3,t)\text{-tree}}}, b_{\mathcal{P}_{(3,t)\text{-tree}}}) = (2, t+1)$. Since every 3-connected graph on F^2 satisfies $(*)$ with $(a,b) = (2, 2g(F^2)-4)$, we expect that such graphs contain a spanning tree T with $\mathrm{te}(T,3) \leq 2g(F^2)-5$. This expectation is indeed true; Kawarabayashi, Nakamoto and Ota [61] proved that every 3-connected locally planar graph G on a surface F^2 contains a spanning 4-tree T with $\mathrm{te}(T,3) \leq 2g(F^2)-5$, and Ozeki [77] proved that every 3-connected graph G on F^2 contains a spanning $\lceil\frac{2g(F^2)+4}{3}\rceil$-tree T with $\mathrm{te}(T,3) \leq 2g(F^2)-5$.

Nakamoto, Oda and Ota [68] considered the existence of a spanning 3-tree with few leaves. Their results are displayed in column $\mathcal{P}_{(2,t)\text{-tree}}$ of Table 3. The bound on t obtained in [68] is almost best possible, but in the planar case, there is a gap. They showed that every 3-connected planar graph contains a spanning 3-tree T with $\mathrm{te}(T,2) \leq \max\left\{\frac{|G|-7}{3}, 0\right\}$ (that is, the number of leaves is at most $\max\left\{\frac{|G|-1}{3}, 2\right\}$), while they conjectured that the best possible value would be $\max\left\{\frac{|G|-11}{3}, 0\right\}$ (that is, the number of leaves could be reduced to at most $\max\left\{\frac{|G|-5}{3}, 2\right\}$).

5.2 Spanning Closed Walks

Recall that we expect that the property of containing a spanning closed k-walk is close to the property of containing a spanning $(k+1)$-tree, while the former is stronger than the latter. With this expectation in mind, almost all results on $\mathcal{P}_{k\text{-tree}}$ in Table 3 were extended to the existence of a spanning closed k-walk. Gao and Richter [43] proved that every 3-connected graph on the sphere or the projective plane contains a spanning closed 2-walk. Cui [28] later gave a short proof to this result. This was extended to 3-connected graphs on the torus and the Klein bottle by Brunet, Ellingham, Gao, Metzlar and Richter [16]. Some results on 3-connected graphs on surfaces were obtained as in Table 3.

As explained in the previous section, Kawarabayashi, Nakamoto and Ota [61] proved that every 3-connected locally planar graph G on a surface F^2 contains a spanning 4-tree T with $\mathrm{te}(T, 3) \leq 2g(F^2) - 5$. They proved a similar result also for a spanning closed 3-walk.

By the above discussion, it is natural to think an extension of the result of the property $\mathcal{P}_{(2,t)\text{-tree}}$ to spanning closed 2-walks, Nakamoto, Oda and Ota [68] implicitly posed the following problem.

Problem 12. *Does there exist a constant s_0 such that every 3-connected planar graph G contains a spanning closed 2-walk in which the number of vertices visited twice is at most $\frac{1}{3}|G| + s_0$?*

This has not been solved yet. Gao, Richter and Yu [44] (see also [45]) showed that every 3-connected planar graph G contains a spanning closed 2-walk in which every vertex visited twice is contained in a 3-cut of G. However, since there exist infinitely many 3-connected planar graphs such that almost all vertices are contained in a 3-cut, this result does not give a non-trivial bound on the number of vertices visited twice in a spanning closed 2-walk.

Note that for a spanning closed 2-walk W, the number of vertices visited twice by W is at most $\frac{1}{3}|G| + s_0$ if and only if the length of W is at most $\frac{4}{3}|G| + s_0$. With this relation in mind, Kawarabayashi and Ozeki [64] proved that every 3-connected planar graph G contains a spanning closed walk of length at most $\frac{4|G|-1}{3}$, but some vertices may be visited three or more times.

5.3 Prism-Hamiltonicity

Another related property, so-called *prism-Hamiltonian*, has been studied. The *prism over a graph* G is defined as the Cartesian product of G and K_2. Thus, it consists of two copies of G and a matching joining the corresponding vertices. A graph G is said to be *prism-Hamiltonian* if the prism over G is Hamiltonian. It is easy to see that any graph with a Hamiltonian path is prism-Hamiltonian, and any prism-Hamiltonian graph contains a spanning closed 2-walk, see [58]. Furthermore, if we denote by $\mathcal{P}_{\text{prism}}$ the property of being prism-Hamiltonian, then we see that $(a_{\mathcal{P}_{\text{prism}}}, b_{\mathcal{P}_{\text{prism}}}) = (a_{\mathcal{P}_{\text{2-walk}}}, b_{\mathcal{P}_{\text{2-walk}}}) = (2, 0)$.

Motivated by the theorem that every 3-connected planar graph contains a spanning closed 2-walk (see Sect. 5.2), Rosenfeld and Barnette [86] (see also [58]) conjectured that every 3-connected planar graph is prism-Hamiltonian, and they proved that any 3-connected cubic planar graph is prism-Hamiltonian[2].

[2] Their proof depends on the Four Color Theorem, which was still unsolved at that time. Without using the Four Color Theorem, the same result was proven by Goodey and Rosenfeld [48] with additional conditions and then finally, by Fleischner [40]. The most general result in this direction is the following due to Paulraja [82]; any 3-connected cubic graph is prism-Hamiltonian.

Biebighauser and Ellingham [9] proved that this conjecture holds for triangulations on the sphere, and also the same conclusion holds for triangulations on the projective plane, the torus, and the Klein bottle. Recently, Spacapan [95] constructed a counterexample to the Rosenfeld-Barnette conjecture. Later, counterexamples with some particuar properties were constructed [54]. On the other hand, Spacapan [96] proved that every 3-connected planar graph with minimum degree at least 4 is prism-Hamiltonian,

6 Related Results

6.1 Graphs Without Complete Bipartite Minors

It is well known that for a 3-connected graph G of order at least 6, G is planar if and only if G does not contain $K_{3,3}$ as a minor (see [67, Lemma 2.5.5], together with Kuratowski's characterization of planar graphs). Considering this characterization, we focus here on 3-connected $K_{3,t}$-minor-free graphs. Note that the lower bound of the genus of the complete bipartite graphs, which is directly obtained by Euler's formula, shows that for any positive integer g, there is an integer t such that any graph on a surface F^2 with $g(F^2) = g$ does not have a $K_{3,t}$-minor[3]. On the other hand, as pointed out in [22], for any g, Böhme, Maharry and Mohar [11] constructed 3-connected $K_{3,7}$-minor-free graphs G that cannot be embedded in an orientable surface F^2 with $g(F^2) = g$. These two results show that studying 3-connected $K_{3,t}$-minor graphs is in fact different from studying 3-connected graphs on surfaces with large Euler genus.

Chen, Egawa, Kawarabayashi, Mohar and Ota [20] showed that every 3-connected $K_{3,t}$-minor-free graph satisfies $(*)$ with $(a, b) = (t-1, -2t+2)$, and for each odd integer t with $t \geq 3$, there exist infinitely many 3-connected $K_{3,t}$-minor-free graphs that satisfy the equality in $(*)$ with $(a, b) = (t-1, -2t+2)$. Recall that $(a_{\mathcal{P}_{k\text{-tree}}}, b_{\mathcal{P}_{k\text{-tree}}}) = (k-1, 1)$. Ota and Ozeki [74] showed the following theorem as expected: for $t \geq 3$, every 3-connected $K_{3,t}$-minor-free graph contains a spanning $(t-1)$-tree when t is even, and a spanning t-tree when t is odd.

Related to this result, it was shown in [35,36] that every 3-connected $K_{2,4}$-minor-free graph and every 3-connected $K_{2,5}$-minor-free planar graph is Hamiltonian, respectively. Using some properties of 3-connected graphs with no $K_{2,t}$-minor, which was shown in [25][4], we can prove that for any integer $t \geq 3$, there exists a constant $c(t)$ such that every 3-connected $K_{2,t}$-minor-free graph contains a spanning tree with at most $c(t)$ leaves. We leave the proof to the readers.

6.2 Hamiltonicity of 4-Connected Graphs with Small Crossing Number

The *crossing number* of a graph G, denoted by $\mathrm{cr}(G)$, is the minimum number of edge crossings over all possible drawings of G on the plane. Note that a

---

[3] This is related to the map color theorem, see [84,85].

[4] Ding [31] gave a complete structure of $K_{2,t}$-minor-free graphs.

graph G is planar if and only if $\mathrm{cr}(G) = 0$. In the same way that Proposition 3 was obtained, we can show that every 4-connected graph G satisfies $(*)$ with $(a, b) = \left(1, \frac{\mathrm{cr}(G)-4}{2}\right)$, which is best possible. With this in mind, recently Ozeki and Zamfirescu [81] conjectured the following:

Conjecture 13 (Ozeki and Zamfirescu [81]**).** *Every 4-connected graph G with $cr(G) \leq 5$ is Hamiltonian.*

Since any graph G with $\mathrm{cr}(G) \leq 1$ has an embedding in the projective plane, Conjecture 13 holds by [98] if we replace the condition $\mathrm{cr}(G) \leq 5$ with $\mathrm{cr}(G) \leq 1$. Ozeki and Zamfirescu [81] extended this to the case of $\mathrm{cr}(G) \leq 2$. Since any graph G with $\mathrm{cr}(G) \leq 2$ can be embedded in the Klein bottle, this is a special case of the open problem stating that every 4-connected graph on the Klein bottle is Hamiltonian, see Table 1. Conjecture 13 is open.

6.3 Hamiltonicity of 3-Connected Planar Graphs

More results on Hamiltonian cycles in planar graphs have been obtained. By Proposition 4, there are some 3-connected planar graphs that do not satisfy $(*)$ with $(a, b) = (1, 0)$, and hence are not Hamiltonian. However, assuming some conditions which guarantee the necessary condition, some results are obtained.

As a corollary of Thomassen's theorem [101], we obtain the result that every plane triangulation with at most one 3-cut is Hamiltonian (which is also shown by Chen [19]). It was shown that the condition "at most one 3-cut" can be replaced with "at most two 3-cuts" (Helden [53]) and also "at most three 3-cuts" (Jackson and Yu [56]). The same conclusion holds for 3-connected planar graphs with at most three 3-cuts (Brinkmann and Zamfirescu [14]). There are 3-connected plane triangulations with six 3-cuts that do not satisfy $(*)$ with $(a, b) = (a, 0)$, and hence are not Hamiltonian (see [13]). Thus, it is natural to conjecture that every 3-connected planar graph with at most five 3-cuts is Hamiltonian. This is still open. Similar work on the properties of containing a Hamiltonian path or of being Hamiltonian-connected can be found in a survey [78].

Dillencourt [30] and Sanders [88] gave another type of sufficient condition for planar graphs to be Hamiltonian.

6.4 Spanning 2-Connected Subgraph with Bounded Degree

A spanning 2-connected subgraph with maximum degree at most k in a graph G is called a *k-covering* (or sometimes a *k-trestle*) of G. While the property of containing a k-covering does not seem to be directly related to the condition $(*)$, this is an extension of a spanning k-tree; some work has been done on this topic. First, Barnette [3] proved that every 3-connected planar graph contains a 15-covering. This result was later improved by Gao [42] to a 6-covering in a 3-connected graph on a surface F^2 with $g(F^2) \leq 2$, which is best possible.

Kawarabayashi, Nakamoto and Ota [60,61] and Sanders and Zhao [90] considered the existence of a k-covering in 3-connected graphs on surfaces with higher genera.

It is also known that with additional conditions, we can reduce the bound on maximum degree in a k-covering. Ellingham and Kawarabayashi [34] showed the existence of a 3-covering in 4-connected locally planar graphs. Enomoto, Iida and Ota [39] proved that every 3-connected planar graph, with minimum degree at least 4, contains a connected subgraph in which every vertex has degree 2 or 3.

Note that Gao and Wormald [46] proved that every plane triangulation contains a spanning closed trail (i.e. a spanning closed walk that does not use an edge twice or more) in which every vertex is visited at most four times. As with a k-covering, this does not seem to be directly related to the condition $(*)$, but this is an extension of a spanning closed walk.

7 Conclusion

In this survey, we introduce a relation between Hamiltonicity and connectivity of graphs on surfaces in terms of the condition $(*)$, which is related to a toughness and scattering number. In fact, we pose the following informal meta-conjecture.

Meta-conjecture. *For any graph property $\mathcal{P}$ related to Hamiltonicity, if all k-connected graphs on a surface F^2 satisfy the condition $(*)$ with $(a, b) = (a_{\mathcal{P}}, b_{\mathcal{P}})$, then every k-connected graph on F^2 satisfies the property $\mathcal{P}$.*

As explained in Sects. 3, 4 and 5, this meta-conjecture holds for several properties $\mathcal{P}$, and in Sect. 6 we show some related results and conjectures that support the meta-conjecture. In this sense, the relation between Hamiltonicity and the connectivity of graphs on surfaces seems to be based on the condition $(*)$.

Acknowledgments. I would thank the referees for carefully reading the paper and helpful comments. This work was supported by JSPS KAKENHI Grant Number JP18K03391.

References

1. Archdeacon, D., Hartsfield, N., Little, C.H.C.: Nonhamiltonian triangulations with large connectivity and representativity. J. Combin. Theory Ser. B **68**, 45–55 (1996)
2. Asano, T., Kikuchi, S., Saito, N.: A linear algorithm for finding Hamiltonian cycles in 4-connected maximal planar graphs. Discrete Appl. Math. **7**, 1–15 (1984)
3. Barnette, D.W.: 2-connected spanning subgraphs of planar 3-connected graphs. J. Combin. Theory Ser. B **61**, 210–216 (1994)
4. Barnette, D.W.: 3-trees in polyhedral maps. Israel J. Math. **79**, 251–256 (1992)
5. Barnette, D.W.: Decomposition theorems for the torus, projective plane and Klein bottle. Discrete Math. **70**, 1–16 (1988)

6. Barnette, D.W.: Trees in polyhedral graphs. Canad. J. Math. **18**, 731–736 (1966)
7. Bauer, D., Broersma, H., Schmeichel, E.: Toughness in graphs - a survey. Graphs Combin. **22**, 1–35 (2006)
8. Bauer, D., Broersma, H., Veldman, H.J.: Not every 2-tough graph is hamiltonian. Discrete Appl. Math. **99**, 317–321 (2000)
9. Biebighauser, D.P., Ellingham, M.N.: Prism-hamiltonicity of triangulations. J. Graph Theory **57**, 181–197 (2008)
10. Biedl, T., Kindermann, P.: Finding Tutte paths in linear time. Preprint (2019). https://arxiv.org/abs/1812.04543
11. Böhme, T., Maharry, J., Mohar, B.: $K_{a,k}$-minors in graphs of bounded tree-width. J. Combin. Theory Ser. B **86**, 135–147 (2002)
12. Böhme, T., Mohar, B., Thomassen, C.: Long cycles in graphs on a fixed surface. J. Combin. Theory Ser. B **85**, 338–347 (2002)
13. Brinkmann, G., Souffriau, J., Van Cleemput, N.: On the strongest form of a theorem of Whitney for Hamiltonian cycles in plane triangulations. J. Graph Theory **83**, 78–91 (2016)
14. Brinkmann, G., Zamfirescu, C.T.: Polyhedra with few 3-cuts are hamiltonian. Electron. J. Combin. **26**, # P1.39 (2019)
15. Broersma, H.J.: On some intriguing problems in hamiltonian graph theory - a survey. Discrete Math. **251**, 47–69 (2002)
16. Brunet, R., Ellingham, M.N., Gao, Z., Metzlar, A., Richter, R.B.: Spanning planar subgraphs of graphs on the torus and Klein bottle. J. Combin. Theory Ser. B **65**, 7–22 (1995)
17. Brunet, R., Nakamoto, A., Negami, S.: Every 5-connected triangulation of the Klein bottle is Hamiltonian. Yokohama Math. J. **47**, 239–244 (1999)
18. Brunet, R., Richter, R.B.: Hamiltonicity of 5-connected toroidal triangulations. J. Graph Theory **20**, 267–286 (1995)
19. Chen, C.: Any maximal planar graph with only one separating triangle is Hamiltonian. J. Comb. Optim. **7**, 79–86 (2003)
20. Chen, G., Egawa, Y., Kawarabayashi, K., Mohar, B., Ota, K.: Toughness of $K_{a,t}$-minor-free graphs, Electron. J. Combin. **18**, # P148 (2011)
21. Chen, G., Fan, G., Yu, X.: Cycles in 4-connected planar graphs. Eur. J. Combin. **25**, 763–780 (2004)
22. Chen, G., Sheppardson, L., Yu, X., Zang, W.: The circumference of a graph with no $K_{3,t}$-minor. J. Combin. Theory Ser. B **96**, 822–845 (2006)
23. Chiba, N., Nishizeki, T.: A theorem on paths in planar graphs. J. Graph Theory **10**, 449–450 (1986)
24. Chiba, N., Nishizeki, T.: The Hamiltonian cycle problem is linear-time solvable for 4-connected planar graphs. J. Algorithms **10**, 187–211 (1989)
25. Chudnovsky, M., Reed, B., Seymour, P.: The edge-density for $K_{2,t}$ minors. J. Combin. Theory Ser. B **101**, 18–46 (2011)
26. Chvátal, V.: Tough graphs and hamiltonian circuits. Discrete Math. **5**, 215–228 (1973)
27. Chvátal, V.: Tough graphs and hamiltonian circuits. Discrete Math. **306**, 910–917 (2006)
28. Cui, Q.: A note on circuit graphs. Electron. J. Discrete Math. **17**, # N10 (2010)
29. Dean, N.: Lecture at Twenty-First Southeastern Conference on Combinatorics, Graph Theory and Computing. Boca Raton, Florida, February 1990
30. Dillencourt, M.B.: Hamiltonian cycles in planar triangulations with no separating triangles. J. Graph Theory **14**, 31–39 (1990)

31. Ding, G.: Graphs without large $K_{2,n}$-minors. Preprint (2020). https://arxiv.org/abs/1702.01355
32. Ellingham, M.N.: Spanning paths, cycles and walks for graphs on surfaces. Congr. Numer. **115**, 55–90 (1996)
33. Ellingham, M.N., Gao, Z.: Spanning trees in locally planar triangulations. J. Combin. Theory Ser. B **61**, 178–198 (1994)
34. Ellingham, M.N., Kawarabayashi, K.: 2-connected spanning subgraphs with low maximum degree in locally planar graphs. J. Combin. Theory Ser. B **97**, 401–412 (2007)
35. Ellingham, M.N., Marshall, E.A., Ozeki, K., Tsuchiya, S.: A characterization of $K_{2,4}$-minor-free graphs. SIAM J. Discrete Math. **30**, 955–975 (2016)
36. Ellingham, M.N., Marshall, E.A., Ozeki, K., Tsuchiya, S.: Hamiltonicity of planar graphs with a forbidden minor. J. Graph Theory **90**, 459–483 (2019)
37. Ellingham, M.N., Zha, X.: Toughness, trees, and walks. J. Graph Theory **33**, 125–137 (2000)
38. Enomoto, H., Jackson, B., Katerinis, P., Saito, A.: Toughness and the existence of k-factors. J. Graph Theory **9**, 87–95 (1985)
39. Enomoto, H., Iida, T., Ota, K.: Connected spanning subgraphs of 3-connected planar graphs. J. Combin. Theory Ser. B **68**, 314–323 (1996)
40. Fleischner, H.: The prism of a 2-connected, planar, cubic graph is hamiltonian. Ann. Discrete Math. **41**, 141–170 (1989)
41. Fujisawa, J., Nakamoto, A., Ozeki, K.: Hamiltonian cycles in bipartite toroidal graphs with a partite set of degree four vertices. J. Combin. Theory Ser. B **103**, 46–60 (2013)
42. Gao, Z.: 2-connected coverings of bounded degree in 3-connected graphs. J. Graph Theory **20**, 327–338 (1995)
43. Gao, Z., Richter, R.B.: 2-walks in circuit graphs. J. Combin. Theory Ser. B **62**, 259–267 (1994)
44. Gao, Z., Richter, R.B., Yu, X.: 2-walks in 3-connected planar graphs. Australas. J. Combin. **11**, 117–122 (1995)
45. Gao, Z., Richter, R.B., Yu, X.: Erratum to: 2-walks in 3-connected planar graphs. Australas. J. Combin. **36**, 315–316 (2006)
46. Gao, Z., Wormald, N.C.: Spanning eulerian subgraphs of bounded degree in triangulations. Graphs Combin. **10**, 123–131 (1994)
47. Goddard, W., Plummer, M.D., Swart, H.C.: Maximum and minimum toughness of graphs of small genus. Discrete Math. **167**(168), 329–339 (1997)
48. Goodey, P.R., Rosenfeld, M.: Hamiltonian circuits in prisms over certain simple 3-polytopes. Discrete Math. **21**, 229–235 (1978)
49. Gouyou-Beauchamps, D.: The Hamiltonian circuit problem is polynomial for 4-connected planar graphs. SIAM J. Comput. **11**, 529–539 (1982)
50. Grünbaum, B.: Polytopes, graphs, and complexes. Bull. Amer. Math. Soc. **76**, 1131–1201 (1970)
51. Harant, J.: Toughness and nonhamiltonicity of polyhedral graphs. Discrete Math. **113**, 249–253 (1993)
52. Hasanvand, M.: Spanning closed walks with bounded maximum degrees of graphs on surfaces. Preprint (2017). https://arxiv.org/abs/1712.00125
53. Helden, G.: Each maximal planar graph with exactly two separating triangles is Hamiltonian. Discrete Appl. Math. **155**, 1833–1836 (2007)
54. Ikegami, D., Maezawa, S., Zamfirescu, C.T.: On 3-polytopes with non-Hamiltonian prisms. J. Graph Theory **97**, 569–577 (2021)

55. Jackson, B., Wormald, N.C.: k-walks of graphs. Australas. J. Combin. **2**, 135–146 (1990)
56. Jackson, B., Yu, X.: Hamilton cycles in plane triangulations. J. Graph Theory **41**, 138–150 (2002)
57. Jung, H.A.: On a class of posets and the corresponding comparability graphs. J. Combin. Theory Ser. B **24**, 125–133 (1978)
58. Kaiser, T., Ryjáček, Z., Král', D., Rosenfeld, M., Voss, H.-J.: Hamilton cycles in prisms. J. Graph Theory **56**, 249–269 (2007)
59. Kawarabayashi, K.: A theorem on paths in locally planar triangulations. Eur. J. Combin. **25**, 781–784 (2004)
60. Kawarabayashi, K., Nakamoto, A., Ota, K.: 2-connected 7-coverings of 3-connected graphs on surfaces. J. Graph Theory **43**, 26–36 (2003)
61. Kawarabayashi, K., Nakamoto, A., Ota, K.: Subgraphs of graphs on surfaces with high representativity. J. Combin. Theory Ser. B **89**, 207–229 (2003)
62. Kawarabayashi, K., Ozeki, K.: 4-connected projective-planar graphs are hamiltonian-connected. J. Combin. Theory Ser. B **112**, 36–69 (2015)
63. Kawarabayashi, K., Ozeki, K.: 5-connected toroidal graphs are Hamiltonian-connected. SIAM J. Discrete Math. **30**, 112–140 (2016)
64. Kawarabayashi, K., Ozeki, K.: Spanning closed walks and TSP in 3-connected planar graphs. J. Combin. Theory Ser. B **109**, 1–33 (2014)
65. Lu, X., West, D.B.: A new proof that 4-connected planar graphs are Hamiltonian-connected. Discuss. Math. Graph Theory **36**, 555–564 (2016)
66. Malkevitch, J.: Polytopal graphs. In: Selected Topics in Graph Theory, vol. 3, pp. 169–188. Academic Press, New York (1988)
67. Mohar, B., Thomassen, C.: Graphs on Surfaces. Johns Hopkins University Press, Baltimore (2001)
68. Nakamoto, A., Oda, Y., Ota, K.: 3-trees with few vertices of degree 3 in circuit graphs. Discrete Math. **309**, 666–672 (2009)
69. Nakamoto, A., Ozeki, K.: Hamiltonian cycles in bipartite quadrangulations on the torus. J. Graph Theory **69**, 143–151 (2012)
70. Nakamoto, A., Ozeki, K.: Spanning trees with few leaves in graphs on surfaces. In: Preparation
71. Nash-Williams, C.S.J.A.: Unexplored and semi-explored territories in graph theory. In: New Directions in the Theory of Graphs, pp. 149–186. Academic Press, New York (1973)
72. Nishizeki, T.: A 1-tough nonhamiltonian maximal planar graph. Discrete Math. **30**, 305–307 (1980)
73. Nishizeki, T., Chiba, N.: Planar Graphs: Theory and Algorithms. Annals of Discrete Mathematics, vol. 32. North-Holland, Amsterdam (1988)
74. Ota, K., Ozeki, K.: Spanning trees in 3-connected $K_{3,t}$-minor-free graphs. J. Combin. Theory Ser. B **102**, 1179–1188 (2012)
75. Owens, P.J.: Non-hamiltonian maximal planar graphs with high toughness. Tatra Mt. Math. Publ. **18**, 89–103 (1999)
76. Ozeki, K.: A shorter proof of Thomassen's theorem on Tutte paths in plane graphs. SUT J. Math. **50**, 417–425 (2014)
77. Ozeki, K.: A spanning tree with bounded maximum degrees of graphs on surfaces. SIAM J. Discrete Math. **27**, 422–435 (2013)
78. Ozeki, K., Van Cleemput, N., Zamfirescu, C.T.: Hamiltonian properties of polyhedra with few 3-cuts - a survey. Discrete Math. **341**, 2646–2660 (2018)
79. Ozeki, K., Vràna, P.: 2-edge-Hamiltonian connectedness of planar graphs. Eur. J. Combin. **35**, 432–448 (2014)

80. Ozeki, K., Yamashita, T.: Spanning trees - a survey. Graphs Combin. **27**, 1–26 (2011)
81. Ozeki, K., Zamfirescu, C.T.: Every 4-connected graph with crossing number 2 is hamiltonian. SIAM J. Discrete Math. **32**, 2783–2794 (2018)
82. Paulraja, P.: A characterization of hamiltonian prisms. J. Graph Theory **17**, 161–171 (1993)
83. Plummer, M.D.: Problem in infinite and finite sets. In: Colloquia Mathematica Societatis János Bolyai, vol. 10, pp. 1549–1559. North-Holland, Amsterdam (1975)
84. Ringel, G.: Das Geschlecht des vollständigen paaren Graphen. Abh. Math. Sem. Univ. Hamburg **28**, 139–150 (1965)
85. Ringel, G.: Der vollständige paare Graph auf nichtorientierbaren Flächen. J. Reine Angew. Math. **220**, 88–93 (1965)
86. Rosenfeld, M.: On spanning subgraphs of 4-connected planar graphs. Graphs Combin. **25**, 279–287 (1989)
87. Rosenfeld, M., Barnette, D.: Hamiltonian circuits in certain prisms. Discrete Math. **5**, 389–394 (1973)
88. Sanders, D.P.: On hamilton cycles in certain planar graphs. J. Graph Theory **21**, 43–50 (1996)
89. Sanders, D.P.: On paths in planar graphs. J. Graph Theory **24**, 341–345 (1997)
90. Sanders, D.P., Zhao, Y.: On 2-connected spanning subgraphs with low maximum degree. J. Combin. Theory Ser. B **74**, 64–86 (1998)
91. Sanders, D.P., Zhao, Y.: On spanning trees and walks of low maximum degree. J. Graph Theory **36**, 67–74 (2001)
92. Schmeichel, E.F., Bloom, G.S.: Connectivity, genus, and the number of components in vertex-deleted subgraphs. J. Combin. Theory Ser. B **27**, 198–201 (1979)
93. Schmid, A., Schmidt, J.M.: Computing 2-walks in polynomial time. ACM Trans. Algorithms **14**, 18 (2018). Art. 22
94. Schmid, A., Schmidt, J.M.: Computing Tutte paths. Preprint (2017). https://arxiv.org/abs/1707.05994
95. Spacapan, S.: A counterexample to prism-hamiltonicity of 3-connected planar graphs. J. Combin. Theory Ser. B **146**, 364–371 (2021)
96. Spacapan, S.: Polyhedra without cubic vertices are prism-hamiltonian. Preprint (2021). https://arxiv.org/abs/2104.04266v1
97. Tait, P.G.: Remarks on the colouring of maps. Proc. R. Soc. London **10**, 729 (1880)
98. Thomas, R., Yu, X.: 4-connected projective-planar graphs are Hamiltonian. J. Combin. Theory Ser. B **62**, 114–132 (1994)
99. Thomas, R., Yu, X.: Five-connected toroidal graphs are hamiltonian. J. Combin. Theory Ser. B **69**, 79–96 (1997)
100. Thomas, R., Yu, X., Zang, W.: Hamilton paths in toroidal graphs. J. Combin. Theory Ser. B **94**, 214–236 (2005)
101. Thomassen, C.: A theorem on paths in planar graphs. J. Graph Theory **7**, 169–176 (1983)
102. Thomassen, C.: Trees in triangulations. J. Combin. Theory Ser. B **60**, 56–62 (1994)
103. Tutte, W.T.: A theorem on planar graphs. Trans. Am. Math. Soc. **82**, 99–116 (1956)
104. Tutte, W.T.: Bridges and Hamiltonian circuits in planar graphs. Aequationes Math. **15**, 1–33 (1977)
105. Tutte, W.T.: On hamiltonian circuits. J. London Math. Soc. **2**, 98–101 (1946)

106. Yu, X.: Disjoint paths, planarizing cycles, and spanning walks. Trans. Am. Math. Soc. **349**, 1333–1358 (1997)
107. Whitney, H.: A theorem on graphs. Ann. Math. **32**, 378–390 (1931)
108. Win, S.: On a connection between the existence of k-trees and the toughness of a graph. Graphs Combin. **5**, 201–205 (1989)

On Structural Parameterizations of Node Kayles

Yasuaki Kobayashi(✉)

Kyoto University, Yoshida-honmachi, Sakyo-ku, Kyoto 606-8501, Japan
kobayashi@iip.ist.i.kyoto-u.ac.jp

Abstract. Node Kayles is a well-known two-player impartial game on graphs: Given an undirected graph, each player alternately chooses a vertex not adjacent to previously chosen vertices, and a player who cannot choose a new vertex loses the game. The problem of deciding if the first player has a winning strategy in this game is known to be PSPACE-complete. A few researches on algorithmic aspects of Node Kayles have been done so far. In this paper, we consider the problem from the view point of fixed-parameter tractability. We show that the problem is fixed-parameter tractable parameterized by the size of a minimum vertex cover or the modular-width of the given graph.

Keywords: Fixed-parameter tractability · Node Kayles · Structural parameterization

1 Introduction

Kayles is a two-player game with bowling pins and a ball. In this game, two players alternately roll a ball down towards a row of pins. Each player knocks down either a pin or two adjacent pins in their turn. The player who knocks down the last pin wins the game. This game has been studied from the viewpoint of combinatorial game theory and the winning player can be characterized by the number of pins at the start of the game.

Schaefer [10] introduced a variant of this game on graphs, which is known as *Node Kayles*. In this game, given an undirected graph, two players alternately choose a vertex, and the vertex chosen and its neighborhood are removed from the graph. The game proceeds as long as the graph has at least one vertex and ends when no vertex is left. The last player wins the game as well as the original game. He studied the computational complexity of this game. In this context, the goal is to decide if the first player can win the game even though the opponent optimally plays the game, that is, the first player has a winning strategy. He showed that this problem (hereinafter referred to simply as Node Kayles) is PSPACE-complete.

After this hardness result was shown, some research on algorithmic aspects of Node Kayles has been done. Bodlaender and Kratsch [3] proved that Node Kayles can be solved in $O(n^{k+2})$ time, where n is the number of vertices and

J. Akiyama et al. (Eds.): JCDCGGG 2018, LNCS 13034, pp. 96–105, 2021.
https://doi.org/10.1007/978-3-030-90048-9_8

k is the asteroidal number of the input graph. This implies several tractability results for some graph classes, including AT-free graphs, circular arc graphs, cocomparability graphs, and cographs since these classes have constant asteroidal numbers. Fleischer and Trippen [7] gave a polynomial-time algorithm for star graphs with arbitrary hair length. Bodlaender et al. [4] studied Node Kayles from the perspective of exact exponential-time algorithms and gave an algorithm that runs in time $O(1.6031^n)$.

We analyze Node Kayles from the view point of parameterized complexity. Here, we consider a *parameterized problem* with instance size n and parameter k. If there is an algorithm that runs in $f(k)n^{O(1)}$ time, where the function f is computable and does not depend on the instance size n, the problem is *fixed-parameter tractable*. There are several possible parameterizations on Node Kayles. One of the most natural ones is the problem of deciding whether the first player has a winning strategy with at most k turns. This problem is known as Short Node Kayles and is known to be AW[*]-complete [1], which means it is unlikely to be fixed-parameter tractable.

For tractable parameterizations, we leverage *structural parameterizations*, meaning that we use the parameters measuring the complexity of graphs rather than that of the problem itself. *Treewidth* is one of the most prominent structural parameterizations for hard graph problems. This parameter measures "tree-likeness" of graphs. In particular, a connected graph has treewidth at most one if and only if it is a tree. Although we know that many graph problems are fixed-parameter tractable when parameterized by the treewidth of the input graph [2], Node Kayles is still open even when the input graph is restricted to trees.

In this paper, we consider two structural parameterizations. We show that Node Kayles is fixed-parameter tractable parameterized by *vertex cover number* or by *modular-width*. More specifically, we show that Node Kayles can be solved in $3^{\tau(G)}n^{O(1)}$ time or $1.6031^{\mathrm{mw}(G)}n^{O(1)}$ time, where $\tau(G)$ is the size of a minimum vertex cover of G and $\mathrm{mw}(G)$ is the modular-width of G. To the best of the author's knowledge, these are the first nontrivial results of the fixed-parameter tractability of Node Kayles.

The algorithm we used in this paper is in fact identical to the one due to Bodlaender et al. [4]. They gave a simple dynamic programming algorithm to solve Node Kayles with the aid of the famous notion *nimber* in the combinatorial game theory. They showed that this dynamic programming algorithm runs in time proportional to the number of specific combinatorial objects, which they call *K-sets*. In this paper, we actually prove that the number of K-sets of G is upper bounded by some functions in $\tau(G)$ or $\mathrm{mw}(G)$, which yields our claims. We also note that our combinatorial analysis is highly stimulated by the work of [8] for the number of minimal separators and potential maximal cliques of graphs.

2 Preliminaries

All graphs appearing in this paper are simple and undirected. Let $G = (V, E)$ be a graph. We denote by $N_G(v)$ the open neighborhood of $v \in V$ in G, that is,

$N_G(v) = \{w \in V : \{v, w\} \in E\}$, and by $N_G[v]$ the closed neighborhood of v in G, that is, $N_G[v] = N_G(v) \cup \{v\}$. Let $X \subseteq V$. We use $G[X]$ to denote the subgraph of G induced by X. We also use the following notations: $N_G(X) = \bigcup_{v \in X} N_G(v) \setminus X$ and $N_G[X] = \bigcup_{v \in X} N_G[v]$.

2.1 Sprague-Grundy Theory

The Sprague-Grundy theory provides unified tools to analyze many two-players impartial combinatorial games. The central idea in this theory is to use *nimber*, which is a nonnegative integer assigned to each position of a game. The nimber of a position can be defined as follows. Hereafter, we consider *normal play games*, that is, the player who makes the last move wins the game. If there is no move from the current position p, the nimber $\mathrm{nim}(p)$ of p is zero. Otherwise, the nimber of p is inductively defined as: $\mathrm{nim}(p) = \mathrm{mex}(\{\mathrm{nim}(p_i) : 1 \leq i \leq m\})$, where $p_1, p_2, \ldots, p_m$ are the positions that can be reached from p with exactly one move, and $\mathrm{mex}(S)$ is the minimum non-negative integer not contained in S. We say that a position is called *a winning position* if the current player has a winning strategy from this position. The following theorem characterizes winning positions of a game with respect to nimbers.

Theorem 1 ([6]). *A position p is a winning position if and only if* $\mathrm{nim}(p) > 0$.

The Sprague-Grundy theory not only allows us a simple way to decide the winning player of a game but also gives an efficient way to compute those nimbers when positions of the game can be decomposed into two or more "independent" subpositions. Consider two positions p_1 and p_2 of (possibly different) games. Then, we can make a new position of the combined game in which each player chooses one of the two positions p_1 and p_2 and then moves from the chosen position to a next position in each turn. When both games are over, so is the combined game. We denote by $p_1 + p_2$ the combined position of p_1 and p_2.

Theorem 2 ([6]). *For any two positions p_1 and p_2, we have* $\mathrm{nim}(p_1 + p_2) = \mathrm{nim}(p_1) \oplus \mathrm{nim}(p_2)$, *where $\oplus$ is the bit-wise exclusive or of the binary representations of the given numbers.*

2.2 K-set

Bodlaender et al. [4] introduced the notion of *K-set* to characterize each game position of Node Kayles.

Definition 1. *A* K-set *is a nonempty subset $W \subseteq V$ of vertices of G such that*

- *$G[W]$ is connected and*
- *there is an independent set $X \subseteq V$ with $W = V \setminus N_G[X]$.*

We call such a triple $(W, N_G(X), X)$ a K-set triple *of G.*

Let us note that a K-set triple partitions V into three sets. Moreover, there are no edges between W and X as $N_G(X)$ separates X from W.

They analyzed Node Kayles via the Sprague-Grundy theorem. In Node Kayles, each position corresponds to some induced graph of the input graph and a single move corresponds to choosing a vertex and then deleting it together with its neighbors. If G has two or more connected components $G_1, G_2, \dots, G_k$, by Theorem 2, the nimber $\mathrm{nim}(G)$ can be computed as $\mathrm{nim}(G_1) \oplus \mathrm{nim}(G_2) \oplus \cdots \oplus \mathrm{nim}(G_k)$. This means that we can independently compute the nimber of each connected component. An important consequence of this fact is that each position, of which we essentially need to compute the nimber, is a subgraph induced by some K-set with chosen vertices X. Thus, a standard recursive dynamic programming algorithm over all K-sets, described in Algorithm 1, solves Node Kayles.

Algorithm 1. A recursive algorithm for computing $\mathrm{nim}(G)$.

Let $G = (V, E)$.
if G is empty **then return** 0 **end if**
if G has two or more connected components $G_1, G_2, \dots, G_k$ **then**
  return $\mathrm{nim}(G_1) \oplus \mathrm{nim}(G_2) \oplus \cdots \oplus \mathrm{nim}(G_k)$
end if
if $\mathrm{nim}(G)$ has been already computed **then**
  return $\mathrm{nim}(G)$
end if
return $\mathrm{mex}(\{\mathrm{nim}(G[V \setminus N_G[v]]) : v \in V\})$

We denote by $\kappa(G)$ the number of K-sets of G.

Lemma 1 ([4]). *Node Kayles can be solved in* $\kappa(G)n^{O(1)}$ *time.*

Bodlaender et al. also gave the following upper bound on the number of K-sets.

Lemma 2 ([4]). *Let G be a graph with n vertices. Then,* $\kappa(G) = O((1.6031 - \varepsilon)^n)$ *for some constant* $\varepsilon > 0$.

These lemmas immediately give an $O(1.6031^n)$-time algorithm for Node Kayles. This raises a natural question: How large is the number of K-sets in general graphs? They also gave the following lower bound example.

Lemma 3 ([4]). *There is a graph G of n vertices satisfying* $\kappa(G) = 3^{(n-1)/3} + 4(n-1)/3$.

Note that $3^{n/3} = \omega(1.4422^n)$ and hence there is still a gap between upper and lower bounds on the number of K-sets. Figure 1 illustrates a lower bound example in Lemma 3. There are four K-sets $\{v_1^i\}, \{v_3^i\}, \{v_2^i, v_3^i\}, \{v_1^i, v_2^i, v_3^i\}$ not containing the root r for each $1 \le i \le (n-1)/3$. Let $P_i = \{v_1^i, v_2^i, v_3^i\}$. For any K-set W containing r, there are three possibilities: $W \cap P_i = \emptyset$, $W \cap P_i = \{v_1^i\}$, or $W \cap P_i = P_i$ for each $1 \le i \le (n-1)/3$. Therefore, there are exactly $3^{(n-1)/3}$ K-sets containing r.

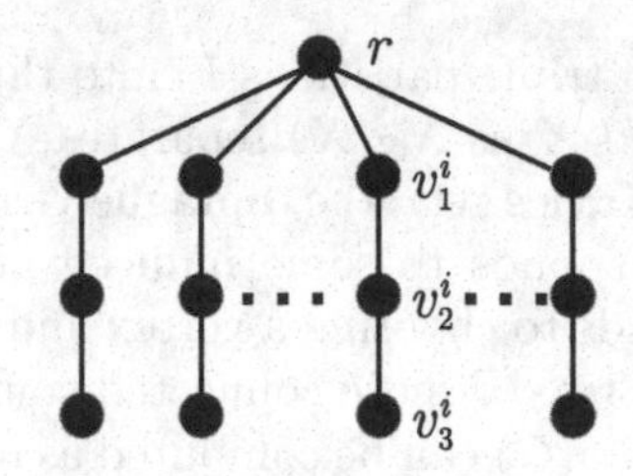

Fig. 1. The lower bound example of Lemma 3.

3 Vertex Cover Number and K-sets

In this section, we give an upper bound on the number of K-sets with respect to the minimum size of a vertex cover of the input graph G. Let $\tau(G)$ be the size of a minimum vertex cover of G.

Theorem 3. $\kappa(G) \leq 3^{\tau(G)} + |V| - \tau(G) - 2^{\tau(G)}$.

Proof. Let C be a vertex cover of G whose size is equal to $\tau(G)$. We say that a tuple (X, Y, Z) is an *ordered tripartition* of C if X, Y, and Z are (possibly empty) disjoint subsets of C such that $X \cup Y \cup Z = C$. From a K-set triple $(W, N_G(X), X)$, we can define an ordered tripartition of C: $(W \cap C, N_G(X) \cap C, X \cap C)$. Conversely, we consider how many K-set triples $(W, N_G(X), X)$ can be obtained from a fixed ordered tripartition (W_C, N_C, X_C) of C, such that $W \cap C = W_C$, $N_G(X) \cap C = N_C$, and $X \cap C = X_C$. First, suppose that W_C is empty. In this case, every K-set W consists of exactly one vertex, which is in $V \setminus C$, and there are only $|V \setminus C|$ possibilities of such K-sets.

Now, suppose that W_C is not empty. We prove that a K-set triple $(W, N_G(X), X)$ is uniquely determined (if it exists) in this case. Obviously, we have $W_C \subseteq W$, $N_C \subseteq N_G(X)$, and $X_C \subseteq X$ for any K-set triple $(W, N_G(X), X)$. Consider a vertex $v \in V \setminus C$ in the independent set. Suppose that v has a neighbor in X_C. Then, we conclude that v is in $N_G(X)$ since X is an independent set and there are no edges between W and X. Next, suppose that v has a neighbor in W_C and has no neighbor in X_C. Then, v must be in W since there are no edges between W and X and every vertex in $N_G(X)$ has a neighbor in X. Suppose otherwise. We conclude that v is in X since $G[W]$ must be connected and v has no neighbor in X.

Therefore, for each ordered tripartition (W_C, N_C, X_C) of C, there is at most one K-set triple $(W, N_G(X), X)$ such that $W \cap C = W_C$, $N_G(X) \cap C = N_C$, and $X \cap C = X_C$, except for the case $W_C = \emptyset$. Clearly, the number of ordered tripartitions (W_C, N_C, X_C) of C with $W_C \neq \emptyset$ is $3^{\tau(G)} - 2^{\tau(G)}$ (as the number of ordered tripartitions with $W_C = \emptyset$ is $2^{\tau(G)}$) and hence the theorem follows. □

It is natural to ask whether the upper bound in Theorem 3 can be improved. However, this is essentially impossible since the lower bound example in Lemma 3 has a minimum vertex cover of size $(n-1)/3+1$, which implies our upper bound is tight up to a constant factor.

4 Modular-Width and K-sets

Let $G = (V, E)$ be a graph. A vertex set $M \subseteq V$ is a *module* of G if for every $v \in V \setminus M$, either $M \cap N_G(v) = \emptyset$ or $M \subseteq N_G(v)$ holds. A *modular decomposition* of G is a recursive decomposition of G into some modules. Here, we define a width parameter associated with this decomposition.

Definition 2. *The* modular-width *of* G*, denoted by* $\mathrm{mw}(G)$*, is the minimum integer* $k \geq 1$ *such that either*

M1 G *consists of a single vertex or*
M2 V *can be partitioned into* $k' \leq k$ *modules* $V_1, V_2, \dots, V_{k'}$ *such that the modular-width of each* $G[V_i]$ *is at most* k *for every* $1 \leq i \leq k'$.

In several papers, the definition of modular-width is slightly different from ours. More specifically, the following two conditions are included in addition to those in Definition 2:

M3 G is a disjoint union of two graphs of modular-width at most k,
M4 G is a join of two graphs of modular-width at most k, that is, G is obtained by taking a disjoint union of the two graphs and adding edges between those graphs.

Note that the conditions M3, M4 are special cases of M2, and this difference may change the modular-width of some extreme cases. However, we emphasize that this does not change our results and then use the simpler version for expository purposes. The modular-width and its decomposition can be computed in linear time [9].

Suppose that G has modular-width at most k. Suppose moreover that V is partitioned into $k' \leq k$ modules $V_1, V_2, \dots, V_{k'}$, where $\mathrm{mw}(G[V_i]) \leq k$ for $1 \leq i \leq k'$. Observe that for every pair of distinct modules V_i and V_j, either $E(V_i, V_j) = \emptyset$ or $E(V_i, V_j) = V_i \times V_j$. To see this, consider an arbitrary $v \in V_i$. Since V_j is a module in G, either $V_j \cap N_G(v) = \emptyset$ or $V_j \cap N_G(v) = V_j$. If $V_j \cap N_G(v) = \emptyset$, $V_j \cap N_G(w) = \emptyset$ holds for every $w \in V_i$ since V_i is also a module in G. Otherwise, $V_j \cap N_G(w) = V_j$ holds for every $w \in V_i$.

Let H be the graph with vertex set $\{v_1, v_2, \dots, v_{k'}\}$ such that v_i is adjacent to v_j if and only if $E(V_i, V_j) = V_i \times V_j$. In other words, H is obtained from G by identifying each module V_i into a single vertex v_i. We say that G is an *expansion graph* of H and, for a subset $U \subseteq \{v_1, v_2, \dots, v_{k'}\}$, the vertex set $\bigcup_{v_i \in U} V_i$ is an *expansion* of U.

Lemma 4. *Let* W *be a K-set of* G *with a K-set triple* $(W, N_G(X), X)$*. Let* H *be a graph with vertex set* $\{v_1, v_2, \dots, v_{k'}\}$*. If* G *is an expansion graph of* H*, then at least one of the following conditions hold:*

- W *is an expansion of some K-set of* H *or*
- W *is a K-set of* $G[V_i]$ *for some* $1 \leq i \leq k'$.

Proof. Observe first that if G has two or more connected components, each module is included in some connected component of G. Moreover, as W is connected, W is also a K-set of a connected component of G. Therefore, in this case, we inductively apply the lemma to this connected component. Thus, we consider the case where G is connected.

We say that two modules V_i and V_j are adjacent if v_i and v_j are adjacent in H. Let $(W, N_G(X), X)$ be a K-set triple of G. In the following, we prove that W is a K-set of $G[V_i]$ for some $1 \leq i \leq k'$ under the assumption that W is not an expansion of any K-sets of H.

We first observe that there is a module V_i with $V_i \cap W \neq \emptyset$ and $V_i \setminus W \neq \emptyset$. To see this, suppose on the contrary, that every module V_i satisfies either $V_i \cap W = \emptyset$ or $V_i \subseteq W$. Let W_H and X_H be the sets of vertices of H whose corresponding modules have an non-empty intersection with W and X, respectively. We claim that $(W_H, N_H(X_H), X_H)$ is a K-set. As $G[W]$ is connected, $H[W_H]$ is connected as well. Similarly, X_H is an independent set of H. By the assumption that either $V_i \cap W = \emptyset$ or $V_i \subseteq W$ for every module V_i, we have $W_H \cap X_H = \emptyset$. If there is an edge between a vertex in W_H and a vertex in X_H, then there are adjacent modules V_i and V_j such that V_i has a vertex in W and V_j has a vertex in X, which contradicts to the fact that $W = N_G[X]$. Thus, $W_H = N_H[X_H]$ and therefore W_H is a K-set of H.

Let V_i be a module of G with $V_i \cap W \neq \emptyset$ and $V_i \setminus W \neq \emptyset$. We then claim that $W \subseteq V_i$. Suppose on the contrary, that there is a module V_j distinct from V_i such that $V_j \cap W \neq \emptyset$. As W is connected in G, we can choose V_j so that V_j is adjacent to V_i. If $V_i \cap X \neq \emptyset$, we have $V_j \subseteq N_G(X)$, which contradicts the fact that V_j has a vertex of W. Moreover, if $V_i \cap N_G(X) \neq \emptyset$, there is an adjacent module V_k with $V_k \cap X \neq \emptyset$ as $V_i \cap X = \emptyset$. This implies that V_i is entirely contained in $N_G(X)$, which also leads to a contradiction. Therefore, we have $W \subseteq V_i$.

Clearly, W is connected in V_i, and $V_i \cap X$ is also an independent set in $G[V_i]$ with $W = V_i \setminus N_{G[V_i]}(V_i \cap X)$. Therefore, W is a K-set in $G[V_i]$. □

For a positive integer n, a *partition of* n is a multiset of positive integers $n_1, n_2, \ldots, n_t$ with $\sum_{1 \leq i \leq t} n_i = n$.

Lemma 5. *For every $\varepsilon > 0$, there is a constant n_ε such that for any integer $n \geq n_\epsilon$ and for any partition $n_1, n_2, \ldots, n_t$ of n with $t \geq 2$, it holds that $\sum_{1 \leq i \leq t} n_i^{1+\varepsilon} + 1 \leq n^{1+\varepsilon}$.*

Proof. Assume that $t = 2$ and $n_1 + n_2 = n$ for some integers n_1, n_2, where $n_1 \geq n_2 \geq 1$. From the convexity of the function $f(x) = x^{1+\varepsilon}$, we have $1 + (n-1)^{1+\varepsilon} \geq n_1^{1+\varepsilon} + n_2^{1+\varepsilon}$. Thus, it suffices to prove the lemma for the case $n_1 = n-1$ and $n_2 = 1$.

Since f is convex and differentiable, by an equivalent characterization of differentiable convex functions [5], we have $f(x) \geq f'(y)(y-x)$ for $x, y \in \mathbb{R}$, which implies

$$f(x) = x^{1+\varepsilon} \geq y^{1+\varepsilon} + (1+\varepsilon)y^{\varepsilon}(x-y)$$

for $x, y \in \mathbb{R}$. Now, we set $x = 1 + 1/(n-1)$ and $y = 1$. Then, it follows that $(1 + 1/(n-1))^{1+\varepsilon} \geq 1 + (1+\varepsilon)/(n-1)$. Thus, we have

$$\begin{aligned} n^{1+\varepsilon} &= (n-1)^{1+\varepsilon}(1 + \frac{1}{n-1})^{1+\varepsilon} \\ &\geq (n-1)^{1+\varepsilon}(1 + \frac{1+\varepsilon}{n-1}) \\ &= (n-1)^{1+\varepsilon} + (1+\varepsilon)(n-1)^{\varepsilon}. \end{aligned}$$

For every $n \geq (\frac{2}{1+\varepsilon})^{1/\varepsilon} + 1$, we have $(1+\varepsilon)(n-1)^{\varepsilon} \geq 2$. Therefore, the lemma holds for $t = 2$. By inductively applying this argument for $t > 2$, the lemma follows. □

Now, we are ready to prove the main claim of this section.

Theorem 4. *For every graph G and every $\varepsilon > 0$, $\kappa(G) = O(1.6031^{\mathrm{mw}(G)} |V|^{1+\varepsilon})$.*

Proof. The proof is by induction on the number of vertices. The base case, where G consists of a single vertex, is trivial. Let $V_1, V_2, \ldots, V_k$ be $k \leq \mathrm{mw}(G)$ modules of G such that $\mathrm{mw}(G[V_i]) \leq \mathrm{mw}(G)$ for each $1 \leq i \leq k$, and let H be a graph of k vertices whose expansion graph is G. By Lemma 4, every K-set of G is either an expansion of a K-set of H or a K-set of some subgraph $G[V_i]$. Let $\mathcal{W}_0$ be the set of K-sets, each of which is an expansion of a K-set of H and let $\mathcal{W}_i$ be the set of K-sets of $G[V_i]$ for $1 \leq i \leq k$. By Lemma 2, there is some constant $c > 0$ such that $|\mathcal{W}_0| = \kappa(H) \leq c \cdot 1.6031^{\mathrm{mw}(G)}$. We apply the induction hypothesis to each $G[V_i]$, and then we have $|\mathcal{W}_i| = \kappa(G[V_i]) \leq c' \cdot 1.6031^{\mathrm{mw}(G)} |V_i|^{1+\varepsilon}$ for some constant $c' > 0$. Summing up all $|\mathcal{W}_i|$, we have

$$\begin{aligned} \kappa(G) &\leq \sum_{0 \leq i \leq k} |\mathcal{W}_i| \\ &\leq \max(c, c') \cdot 1.6031^{\mathrm{mw}(G)} (1 + \sum_{1 \leq i \leq k} |V_i|^{1+\varepsilon}) \\ &= O(1.6031^{\mathrm{mw}(G)} |V|^{1+\varepsilon}). \end{aligned}$$

The last equality follows from Lemma 5. □

Let us note that this argument also works for some graph classes. In particular, Bodlaender et al. [4] proved that $\kappa(T) \leq n \cdot 3^{n/3}$ for any tree with n vertices. For any modular decomposition of a tree T, the graph H whose expansion graph is T induces a tree. From this fact, we immediately have $\kappa(T) = 3^{\mathrm{mw}(T)/3} n^{1+\varepsilon}$.

5 Concluding Remarks

In this paper, we give a new running time analysis of the known algorithm for Node Kayles. Bodlaender et al. [4] showed that Node Kayles can be solved in

$\kappa(G)n^{O(1)}$ time, where $\kappa(G)$ is the number of K-sets in G and that $\kappa(G) = O(1.6031^n)$. We analyze the number of K-sets from the perspective of structural parameterizations of graphs, and show that $\kappa(G) \leq 3^{\tau(G)} + n - \tau(G) - 2^{\tau(G)}$ and $\kappa(G) = O(1.6031^{\mathrm{mw}(G)} n^{1+\varepsilon})$ for every constant $\varepsilon > 0$. The first upper bound is tight up to the constant factor, and the second one improves the known upper bound due to Bodlaender et al. [4] when the modular-width is relatively small compared to the number of vertices.

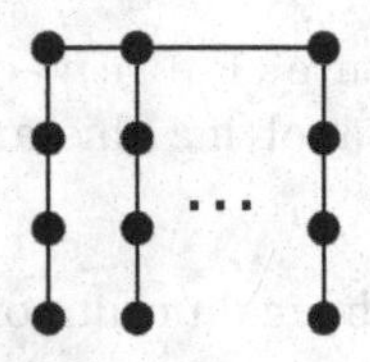

Fig. 2. Similar to the example in Lemma 3, this graph has exponentially many K-sets.

It would be interesting to know whether other graph parameters yield a new upper bound on the number of K-sets. However, the lower bound example in Lemma 3 indicates some limitation on this question. In particular, the treewidth, pathwidth, and even treedepth of this example are all bounded. Moreover, we can transform the lower bound example into a bounded degree tree as in Fig. 2, which has also exponentially many K-sets. Note that this argument does not imply the hardness of Node Kayles on trees.

Acknowledgments. The author thanks Kensuke Kojima for simplifying the proof of Lemma 5 and anonymous reviewers for valuable comments. This work was partially supported by JST CREST JPMJCR1401.

References

1. Abrahamson, K.R., Downey, R.G., Fellows, M.R.: Fixed-parameter tractability and completeness IV: on completeness for W[P] and PSPACE analogues. Ann. Pure Appl. Log. **73**(3), 235–276 (1995). https://doi.org/10.1016/0168-0072(94)00034-Z
2. Bodlaender, H.L.: A tourist guide through treewidth. Acta Cybern. **11**(1–2), 1–21 (1993)
3. Bodlaender, H.L., Kratsch, D.: Kayles and nimbers. J. Algorithms **43**(1), 106–119 (2002). https://doi.org/10.1006/jagm.2002.1215
4. Bodlaender, H.L., Kratsch, D., Timmer, S.T.: Exact algorithms for Kayles. Theor. Comput. Sci. **562**, 165–176 (2015). https://doi.org/10.1016/j.tcs.2014.09.042
5. Boyd, S., Vandenberghe, L.: Convex Optimization. Cambridge University Press, Cambridge (2004)
6. Conway, J.H.: On Numbers and Games, 2nd edn. A K Peters, Natick (2001)
7. Fleischer, R., Trippen, G.: Kayles on the way to the stars. In: van den Herik, H.J., Björnsson, Y., Netanyahu, N.S. (eds.) CG 2004. LNCS, vol. 3846, pp. 232–245. Springer, Heidelberg (2006). https://doi.org/10.1007/11674399_16

8. Fomin, F.V., Liedloff, M., Montealegre, P., Todinca, I.: Algorithms parameterized by vertex cover and modular width, through potential maximal cliques. Algorithmica **80**(4), 1146–1169 (2018)
9. McConnell, R.M., Spinrad, J.P.: Linear-time modular decomposition and efficient transitive orientation of comparability graphs. In: SODA, pp. 536–545. ACM/SIAM (1994)
10. Schaefer, T.J.: On the complexity of some two-person perfect-information games. J. Comput. Syst. Sci. **16**(2), 185–225 (1978)

Robustness in Power-Law Kinetic Systems with Reactant-Determined Interactions

Noel T. Fortun[1(✉)], Angelyn R. Lao[1], Luis F. Razon[2], and Eduardo R. Mendoza[1,3,4,5]

[1] Mathematics and Statistics Department, De La Salle University, 0922 Manila, Philippines
noel.fortun@dlsu.edu.ph

[2] Chemical Engineering Department, De La Salle University, 0922 Manila, Philippines

[3] Institute of Mathematical Sciences and Physics, University of the Philippines, 4031 Los Baños, Laguna, Philippines

[4] Max Planck Institute of Biochemistry, Martinsried near Munich, Germany

[5] Faculty of Physics, Ludwig Maximilian University, 80539 Munich, Germany

Abstract. Robustness against the presence of environmental disruptions can be observed in many systems of chemical reaction network. However, identifying the underlying components of a system that give rise to robustness is often elusive. The influential work of Shinar and Feinberg established simple yet subtle network-based conditions for absolute concentration robustness (ACR), a phenomenon in which a species in a mass-action system has the same concentration for any positive steady state the network may admit. In this contribution, we extend this result to embrace kinetic systems more general than mass-action systems, namely, power-law kinetic systems with reactant-determined interactions (denoted by "PL-RDK"). In PL-RDK, the kinetic order vectors of reactions with the same reactant complex are identical. As illustration, we considered a scenario in the pre-industrial state of global carbon cycle. A power-law approximation of the dynamical system of this scenario is found to be dynamically equivalent to an ACR-possessing PL-RDK system.

Keywords: Absolute concentration robustness · Chemical reaction network · Power-law kinetics · Reactant-determined interactions · Carbon cycle model

1 Introduction

Robustness may be generally defined [15,26] as a system-level dynamical property that allows a system to sustain its functions despite changes in internal and external conditions. This feature, in fact, is fundamental and ubiquitous

J. Akiyama et al. (Eds.): JCDCGGG 2018, LNCS 13034, pp. 106–121, 2021.
https://doi.org/10.1007/978-3-030-90048-9_9

in many biological processes, including cellular networks and entire organisms [2,15,26]. One type of robust behavior is "concentration robustness," wherein some quantity involving the concentrations of the different species in a network is fixed at equilibrium [10]. In a well-cited paper published in *Science*, Shinar and Feinberg [26] introduced *absolute concentration robustness* (ACR), a condition in which the concentration of a species in a network attains the same value in every positive steady state set by parameters and does not depend on initial conditions.

Shinar and Feinberg presented sufficient structure-based conditions for a chemical reaction network (CRN) to display ACR on a particular species through a structural index called the *deficiency*. This non-negative parameter has been the center of many powerful results in *Chemical Reaction Network Theory* (CRNT), a theoretical body of work that associates the structure of a CRN to the dynamical behaviour of the system [11,12]. CRNT employs mathematical methods from graph theory, linear algebra, group theory and the theory of ordinary differential equations. In CRNT, chemical reaction networks are viewed as digraphs whose vertices (called *complexes*) are mapped to non-negative vectors representing compositions of chemical species and whose arcs represent chemical reactions between them. The Shinar-Feinberg Theorem on ACR holds for systems whose evolution are modelled by ordinary differential equations with mass-action kinetics (MAK), and is stated as follows:

Consider a mass-action system that admits a positive steady state and suppose that the deficiency of the underlying reaction network is one. If there are two nonterminal nodes in the network that differ only in species S, then the system has absolute concentration robustness in S.

Here, we show that this result extends to systems endowed with *power-law kinetics* (PLK), which generalize mass-action kinetics [7,14]. Several experiments have shown that the kinetic order of a reaction with respect to a given reactant is a function of the geometry within which the reaction occurs [16–18,22,25]. In the case of reactions occurring within a three-dimensional homogenous space (as in mass-action systems), the kinetic order is the same as the number of molecules entering into the reaction. However, for systems characterized by molecular overcrowding (e.g., when other molecules deny the reactants from the supposedly allowable space, and to stickiness, when the reactants are found along the surfaces of the reaction vessel) the kinetic orders for the reactions can exhibit non-integer values [24] found in power-law formalism [23,28,29]. For instance, in intracellular environments, which are highly structured and characterized by molecular crowding, reactions in vivo are likely to take place on membranes or channels and as such, reactions follow fractal-like kinetics [6,8,19,25]. The presence of power-law kinetics in reaction systems thus motivated CRN-based studies on PLK systems ([9,13,20,27] among others), some of which are extensions or modifications of existing results on MAK systems.

This contribution specifically shows that the result of Shinar and Feinberg on ACR applies to a class of PLK system called *power-law kinetic systems with reactant-determined interactions* (denoted by "PL-RDK"). PL-RDK systems are kinetic systems with power-law rate functions whose kinetic order vectors are identical for reactions with the same reactant complex. Since the kinetic orders of the mass-action rate functions are precisely the stoichiometric coefficients of the reactant complex, one can see that MAK is a special case of PL-RDK.

As an application, we employ the theorem to a power-law approximation of the ODE system corresponding to a specific scenario in the pre-industrial carbon cycle model developed by Anderies et al. [1]. Particularly, for the pre-industrial scenario where there are anthropogenic causes that reduce the capacity of terrestrial carbon pool to store carbon, the power-law approximation leads to an ACR-possessing PL-RDK system.

The rest of the paper is organized as follows: Sect. 2 assembles preliminary concepts in Chemical Reaction Network Theory required in stating and proving the results. Section 3 discusses the extension of the Shinar-Feinberg Theorem on ACR for PL-RDK systems. Section 4 applies the main result obtained from the previous section to a carbon cycle model. In Sect. 5, we summarize our results and outline some research perspectives.

2 Fundamentals of Chemical Reaction Networks and Kinetic Systems

We recall some fundamental notions about chemical reaction networks (CRNs) and chemical kinetic systems (CKS) assembled in [5,27]. Some concepts introduced by Feinberg in [11,12] are also reviewed.

Notation. We denote the real numbers by $\mathbb{R}$, the non-negative real numbers by $\mathbb{R}_{\geq 0}$ and the positive real numbers by $\mathbb{R}_{>0}$. Objects in the reaction systems are viewed as members of vector spaces. Suppose $\mathscr{I}$ is a finite index set. By $\mathbb{R}^{\mathscr{I}}$, we mean the usual vector space of real-valued functions with domain $\mathscr{I}$. For $x \in \mathbb{R}^{\mathscr{I}}$, the ith coordinate of x is denoted by x_i, where $i \in \mathscr{I}$. The sets $\mathbb{R}^{\mathscr{I}}_{\geq 0}$ and $\mathbb{R}^{\mathscr{I}}_{>0}$ are called the *non-negative* and *positive orthants* of $\mathbb{R}^{\mathscr{I}}$, respectively. Addition, subtraction, and scalar multiplication in $\mathbb{R}^{\mathscr{I}}$ are defined in the usual way. If $x \in \mathbb{R}^{\mathscr{I}}_{>0}$ and $y \in \mathbb{R}^{\mathscr{I}}$, we define $x^y \in \mathbb{R}_{>0}$ by $x^y = \prod_{i \in \mathscr{I}} x_i^{y_i}$. The vector $\log x \in \mathbb{R}^{\mathscr{I}}$, where $x \in \mathbb{R}^{\mathscr{I}}_{>0}$, is given by $(\log x)_i = \log x_i$, for all $i \in \mathscr{I}$. If $x, y \in \mathbb{R}^{\mathscr{I}}$, the standard scalar product $x \cdot y \in \mathbb{R}$ is defined by $x \cdot y = \sum_{i \in \mathscr{I}} x_i y_i$. By the *support* of $x \in \mathbb{R}^{\mathscr{I}}$, denoted by supp x, we mean the subset of $\mathscr{I}$ assigned with non-zero values by x. That is, supp $x := \{i \in \mathscr{I} | x_i \neq 0\}$.

Definition 1. *A* ***chemical reaction network*** *(CRN) $\mathscr{N}$ is a triple $(\mathscr{S}, \mathscr{C}, \mathscr{R})$ of three finite sets:*

1. *a set $\mathscr{S} = \{X_1, X_2, \ldots, X_m\}$ of* ***species****;*
2. *a set $\mathscr{C} \subset \mathbb{R}^{\mathscr{S}}_{\geq 0}$ of* ***complexes****;*
3. *a set $\mathscr{R} = \{R_1, R_2, \ldots, R_r\} \subset \mathscr{C} \times \mathscr{C}$ of* ***reactions*** *such that $(y, y) \notin \mathscr{R}$ for any $y \in \mathscr{C}$, and for each $y \in \mathscr{C}$, there exists $y' \in \mathscr{C}$ such that either $(y, y') \in \mathscr{R}$ or $(y', y) \in \mathscr{R}$.*

We denote the number of species with m, the number of complexes with n and the number of reactions with r.

A CRN can be viewed as a digraph $(\mathscr{C}, \mathscr{R})$ with vertex-labelling. In particular, it is a digraph where each vertex $y \in \mathscr{C}$ has positive degree and stoichiometry, i.e., there is a finite set $\mathscr{S}$ of species such that $\mathscr{C}$ is a subset of $\mathbb{R}^{\mathscr{S}}_{\geq 0}$. The vertices are the complexes whose coordinates are in $\mathbb{R}^{\mathscr{S}}_{\geq 0}$, which are the **stoichiometric coefficients**. The arcs are precisely the reactions.

We use the convention that an element $R_j = (y_j, y_j') \in \mathscr{R}$ is denoted by $R_j : y_j \to y_j'$. In this reaction, we say that y_j is the **reactant** complex and y_j' is the **product** complex. Connected components of a CRN are called **linkage classes**, strongly connected components are called **strong linkage classes**, and strongly connected components without outgoing arcs are called **terminal strong linkage classes**. We denote the number of linkage classes with ℓ, that of the strong linkage classes with $s\ell$, and that of terminal strong linkage classes with t. A complex is called **terminal** if it belongs to a terminal strong linkage class; otherwise, the complex is called **nonterminal**.

With each reaction $y \to y'$, we associate a **reaction vector** obtained by subtracting the reactant complex y from the product complex y'. The **stoichiometric subspace** S of a CRN is the linear subspace of $\mathbb{R}^{\mathscr{S}}$ defined by

$$S := \text{span}\,\{y' - y \in \mathbb{R}^{\mathscr{S}}\,| y \to y' \in \mathscr{R}\}.$$

The **rank** of the CRN, s, is defined as $s = \dim S$.

Many features of CRNs can be examined by working in terms of finite dimensional spaces $\mathbb{R}^{\mathscr{S}}$ (species space), $\mathbb{R}^{\mathscr{C}}$ (complex space), and $\mathbb{R}^{\mathscr{R}}$ (reaction space). Suppose the set $\{\omega_i \in \mathbb{R}^{\mathscr{I}} \mid i \in \mathscr{I}\}$ forms the *standard basis* for $\mathbb{R}^{\mathscr{I}}$ where $\mathscr{I} = \mathscr{S}, \mathscr{C}$ or $\mathscr{R}$. We recall four maps relevant in the study of CRNs: map of complexes, incidence map, stoichiometric map and Laplacian map.

Definition 2. *Let $\mathscr{N} = (\mathscr{S}, \mathscr{C}, \mathscr{R})$ be a CRN.*

1. *The* ***map of complexes*** $Y : \mathbb{R}^{\mathscr{C}} \to \mathbb{R}^{\mathscr{S}}$ *maps the basis vector ω_y to the complex $y \in \mathscr{C}$.*
2. *The* ***incidence map*** $I_a : \mathbb{R}^{\mathscr{R}} \to \mathbb{R}^{\mathscr{C}}$ *is the linear map defined by mapping for each reaction $R_j : y_j \to y_j' \in \mathscr{R}$, the basis vector ω_j to the vector $\omega_{y_j'} - \omega_{y_j} \in \mathscr{C}$.*
3. *The* ***stoichiometric map*** $N : \mathbb{R}^{\mathscr{R}} \to \mathbb{R}^{\mathscr{S}}$ *is defined as $N = Y \circ I_a$.*
4. *For each $k \in \mathbb{R}^{\mathscr{R}}_{>0}$, the linear transformation $A_k : \mathbb{R}^{\mathscr{C}} \to \mathbb{R}^{\mathscr{C}}$ called* ***Laplacian map*** *is the mapping defined by*

$$A_k x := \sum_{y \to y' \in \mathscr{R}} k_{y \to y'} x_y (\omega_{y'} - \omega_y),$$

where x_y refers to the yth component of $x \in \mathbb{R}^{\mathscr{C}}$ relative to the standard basis.

The following result, named as the *Structure Theorem of the Laplacian Kernel* (STLK) by Arceo et al. in [5], is crucial in deriving important results in CRNT [11,12].

Proposition 1 (Structure Theorem of the Laplacian Kernel (STLK), Prop. 4.1 [11]). *Let $\mathscr{N} = (\mathscr{S}, \mathscr{C}, \mathscr{R})$ be a CRN with terminal strong linkage classes $\mathscr{C}^1, \mathscr{C}^2, \ldots, \mathscr{C}^t$. Let $k \in \mathbb{R}^{\mathscr{R}}_{>0}$ and A_k its associated Laplacian. Then Ker A_k has a basis $b^1, b^2, \ldots, b^t$ such that supp $b^i = \mathscr{C}^i$ for all $i = 1, 2, \ldots, t$.*

A non-negative integer, called the deficiency, can be associated to each CRN. The **deficiency** of a CRN, denoted by δ, is the integer defined by $\delta = n - \ell - s$. This index has been the center of many studies in CRNT due to its relevance in the dynamic behaviour of the system. In [11], Feinberg provided a geometric interpretation of deficiency: $\delta = \dim(\text{Ker } Y \cap \text{Im } I_a)$. This fact and the STLK imply the following corollary.

Corollary 1 (Cor. 4.12 [11]). *Let $\mathscr{N} = (\mathscr{S}, \mathscr{C}, \mathscr{R})$ be a CRN with deficiency δ and t terminal strong linkage classes. Then for each $k \in \mathbb{R}^{\mathscr{R}}_{>0}$,*

$$\dim(\mathit{Ker}\, Y A_k) \leq \delta + t.$$

By *kinetics* of a CRN, we mean the assignment of a rate function to each reaction in the CRN. It is defined formally as follows.

Definition 3. *A* ***kinetics*** *of a CRN $\mathscr{N} = (\mathscr{S}, \mathscr{C}, \mathscr{R})$ is an assignment of a rate function $K_j : \Omega_K \to \mathbb{R}_{\geq 0}$ to each reaction $R_j \in \mathscr{R}$, where Ω_K is a set such that $\mathbb{R}^{\mathscr{S}}_{>0} \subseteq \Omega_K \subseteq \mathbb{R}^{\mathscr{S}}_{\geq 0}$. A kinetics for a network $\mathscr{N}$ is denoted by*

$$K = [K_1, K_2, ..., K_r]^\top : \Omega_K \to \mathbb{R}^{\mathscr{R}}_{\geq 0}.$$

The pair $(\mathscr{N}, K)$ is called the ***chemical kinetic system (CKS)****.*

The above definition is adopted from [30]. It is expressed in a more general context than what one typically finds in CRNT literature. For power-law kinetic systems, one sets $\Omega_K = \mathbb{R}^{\mathscr{S}}_{>0}$. Here, we focus on the kind of kinetics relevant to our context:

Definition 4. *A* ***chemical kinetics*** *is a kinetics K satisfying the positivity condition:*

For each reaction $R_j : y_j \to y'_j \in \mathscr{R}$, $K_j(c) > 0$ if and only if supp $y_j \subset$ supp c.

Once a kinetics is associated with a CRN, we can determine the rate at which the concentration of each species evolves at composition $c \in \mathbb{R}^{\mathscr{S}}_{>0}$.

Definition 5. *The* ***species formation rate function*** *of a chemical kinetic system is the vector field*

$$f(c) = NK(c) = \sum_{y_j \to y'_j \in \mathscr{R}} K_j(c)(y'_j - y_j).$$

The equation $dc/dt = f(c)$ is the ***ODE or dynamical system*** *of the CKS. A* ***positive equilibrium or steady state*** *c^* is an element of $\mathbb{R}^{\mathscr{S}}_{>0}$ for which $f(c^*) = 0$. The set of positive equilibria of a chemical kinetic system is denoted by $E_+(\mathscr{N}, K)$.*

Power-law kinetics is defined by an $r \times m$ matrix $F = [F_{ij}]$, called the **kinetic order matrix**, and vector $k \in \mathbb{R}^{\mathscr{R}}$, called the **rate vector**.

Definition 6. *A kinetics* $K : \mathbb{R}^{\mathscr{S}}_{>0} \to \mathbb{R}^{\mathscr{R}}$ *is a* ***power-law kinetics*** *(PLK) if*

$$K_i(x) = k_i x^{F_{i,\cdot}} \quad \forall i = 1, \ldots, r$$

with $k_i \in \mathbb{R}_{>0}$ *and* $F_{ij} \in \mathbb{R}$. *A PLK system has* ***reactant-determined kinetics*** *(of type* ***PL-RDK****) if for any two reactions* R_i, $R_j \in \mathscr{R}$ *with identical reactant complexes, the corresponding rows of kinetic orders in* F *are identical, i.e.,* $F_{ik} = F_{jk}$ *for* $k = 1, ..., m$.

An example of PL-RDK is the well-known **mass-action kinetics** (MAK), where the kinetic order matrix is the transpose of the matrix representation of the map of complexes Y [11]. That is, a kinetics is a MAK if

$$K_j(c) = k_j x^{Y_{\cdot,j}} \quad \text{for all } R_j : y_j \to y_j' \in \mathscr{R}$$

where $k_j \in \mathbb{R}_{>0}$, called rate constants. Note that $Y_{\cdot,j}$ pertains to the stoichiometric coefficients of a reactant complex $y_j \in \mathscr{C}$.

Remark 1. In [5], Arceo et al. discussed several sets of kinetics of a network and drew a "kinetic landscape". They identified two main sets: the *complex factorizable* kinetics and its complement, the *non-complex factorizable* kinetics. Complex factorizable kinetics generalize the key structural property of MAK – that is, the species formation rate function decomposes as

$$\frac{dx}{dt} = Y \circ A_k \circ \Psi_k,$$

where Y is the map of complexes, A_k is the Laplacian map, and $\Psi_k : \mathbb{R}^{\mathscr{S}}_{\geq 0} \to \mathbb{R}^{\mathscr{C}}_{\geq 0}$ such that $I_a \circ K(x) = A_k \circ \Psi_k(x)$ for all $x \in \mathbb{R}^{\mathscr{S}}_{\geq 0}$. In the set of power-law kinetics, PL-RDK is the subset of complex-factorizable kinetics.

We recall the definition of the $m \times n$ matrix $\widetilde{Y}$ from the work of Müller and Regensburger [20,21]: For each reactant complex, the associated column of $\widetilde{Y}$ is the transpose of the kinetic order matrix row of the complex's reaction, otherwise (i.e., for non-reactant complexes), the column is 0. We form the ***T*-matrix** of a PL-RDK system by truncating away the columns of the non-reactant complexes in $\widetilde{Y}$, obtaining an $m \times n_r$ matrix, where n_r denotes the number of reactant complexes [27].

3 Absolute Concentration Robustness in PL-RDK Systems

To illustrate absolute concentration robustness, we consider the following toy model:

(1)

The map depicts a biochemical system involving transfer of material from two pools: X_2 to X_1 and X_1 to X_2, but with X_2 regulating the second process. Suppose the system evolves according to the following set of ODEs:

$$\begin{aligned} \dot{X}_1 &= k_1 X_2^{0.8} - k_2 X_1^{0.5} X_2^{0.8} \\ \dot{X}_2 &= -k_1 X_2^{0.8} + k_2 X_1^{0.5} X_2^{0.8} \end{aligned} \tag{2}$$

The positive equilibrium of the system is attained when

$$X_1 = \left(\frac{k_1}{k_2}\right)^2 \quad \text{and} \quad X_2 = \Gamma - \left(\frac{k_1}{k_2}\right)^2. \tag{3}$$

where Γ is the conserved amount of total material. These equations indicate that whenever $\Gamma > (k_1/k_2)^2$, a positive steady state exists. Furthermore, since X_1 has the same value in any steady state, the system exhibits ACR in X_1.

We define absolute concentration robustness in PL-RDK systems as follows:

Definition 7. *A PL-RDK system $(\mathscr{N}, K)$ has* ***absolute concentration robustness(ACR) in species*** $X_i \in \mathscr{S}$ *if there exists $c^* \in E_+(\mathscr{N}, K)$ and for every other $c^{**} \in E_+(\mathscr{N}, K)$, we have $c_i^{**} = c_i^*$.*

The following proposition adapts Theorem S3.15 found in supplementary online material of the paper of Shinar and Feinberg [26] to deal with PL-RDK systems.

Proposition 2. *Let $\mathscr{N} = (\mathscr{S}, \mathscr{C}, \mathscr{R})$ be a deficiency-one CRN. Suppose that $(\mathscr{N}, K)$ is a PL-RDK system which admits a positive equilibrium c^*. If $y, y' \in \mathscr{C}$ are nonterminal complexes, then each positive equilibrium c^{**} of the system satisfies the equation*

$$(T_{\cdot,y} - T_{\cdot,y'}) \cdot \log\left(\frac{c^{**}}{c^*}\right) = 0. \tag{4}$$

We largely reproduce the proof of Shinar and Feinberg in the said supplementary material of their paper. Since in their proof, the sums are often taken over all complexes, we use the notation of Müller and Regensburger in [20, 21]:

$$\widetilde{Y} = [\ T \mid 0\ ],$$

adjoining $n - n_r$ zero columns for the non-reactant complexes, where n_r denotes the number of reactant complexes. Furthermore, we write $\widetilde{y}$ for $\widetilde{Y}_{\cdot,y}$.

Proof. Assume that c^* is a positive steady state of the PL-RDK system $(\mathscr{N}, K)$. That is,

$$\sum_{y \to y' \in \mathscr{R}} k_{y \to y'} (c^*)^{\widetilde{y}} (y' - y) = 0. \tag{5}$$

For each $y \to y' \in \mathscr{R}$, define the positive number $\kappa_{y \to y'}$ by

$$\kappa_{y \to y'} := k_{y \to y'} (c^*)^{\widetilde{y}}. \tag{6}$$

Thus, we obtain

$$\sum_{y \to y' \in \mathscr{R}} \kappa_{y \to y'} (y' - y) = 0. \tag{7}$$

Suppose that c^{**} is also a positive equilibrium of the system. Hence,

$$\sum_{y \to y' \in \mathscr{R}} k_{y \to y'} (c^{**})^{\widetilde{y}} (y' - y) = 0. \tag{8}$$

Define

$$\mu := \log c^{**} - \log c^*. \tag{9}$$

With $\kappa \in \mathbb{R}^{\mathscr{R}}_{>0}$ given by Eq. (6) and μ given by Eq. (9), it follows from Eq. (8) that

$$\sum_{y \to y' \in \mathscr{R}} \kappa_{y \to y'} e^{\widetilde{y} \cdot \mu} (y' - y) = 0. \tag{10}$$

Let $\mathbf{1}^{\mathscr{C}} \in \mathbb{R}^{\mathscr{C}}$ such that

$$\mathbf{1}^{\mathscr{C}} = \sum_{y \in \mathscr{C}} \omega_y.$$

Observe that Eqs. (7) and (10) can be respectively written as

$$Y A_\kappa \mathbf{1}^{\mathscr{C}} = 0, \text{ and } Y A_\kappa \left(\sum_{y \in \mathscr{C}} e^{\widetilde{y} \cdot \mu} \omega_y \right) = 0.$$

Equivalently,

$$\mathbf{1}^{\mathscr{C}} \in \text{Ker } Y A_\kappa, \text{ and} \tag{11}$$

$$\sum_{y \in \mathscr{C}} e^{\widetilde{y} \cdot \mu} \omega_y \in \text{Ker } Y A_\kappa. \tag{12}$$

Therefore, c^* and c^{**} are positive equilibria of the PL-RDK system $(\mathscr{N}, K)$ if and only if Eqs. (11) and (12) hold. From Corollary 1, we have

$$\dim(\text{Ker } Y A_\kappa) \leq 1 + t \tag{13}$$

for the CRN under consideration. Let $\{b^1, b^2, \ldots, b^t\} \subset \mathbb{R}^{\mathscr{C}}_{\geq 0}$ be a basis for Ker A_κ as in Proposition 1 (STLK). Since Ker $A_\kappa \subseteq$ Ker $Y A_\kappa$, this basis of Ker A_κ can be extended to form a basis of Ker $Y A_\kappa$. Recall from Eq. (11) that $\mathbf{1}^{\mathscr{C}}$ is in

$\text{Ker}\, YA_\kappa$. We assert that the set $\{\mathbf{1}^{\mathscr{C}}, b^1, b^2, \ldots, b^t\}$ is a basis for $\text{Ker}\, YA_\kappa$ (and hence, equality holds in Eq. (13)). This follows if

$$\mathbf{1}^{\mathscr{C}} \notin \text{Span}\, \{b^1, b^2, \ldots, b^t\}. \tag{14}$$

From Proposition 1, every element of $\text{Ker}\, A_\kappa$ must have its support contained entirely in the set of terminal complexes. However, the support of $\mathbf{1}^{\mathscr{C}}$ consists of all complexes. By assumption, there are nonterminal complexes and hence, $\mathbf{1}^{\mathscr{C}}$ cannot lie in $\text{Ker}\, A_\kappa$ (i.e., Eq. (14) holds).

From Eq. (12), there exist scalars $\lambda_0, \lambda_1, \ldots, \lambda_t$ such that

$$\sum_{y \in \mathscr{C}} e^{\widetilde{y} \cdot \mu} \omega_y = \lambda_0 \mathbf{1}^{\mathscr{C}} + \sum_{i=1}^{t} \lambda_i b^i. \tag{15}$$

Observe that each vector b^i, $i = 0, 1, \ldots, t$, has its support entirely on terminal complexes. This fact, along with Eq. (15), implies that for each pair of nonterminal complexes $y \in \mathscr{C}$ and $y' \in \mathscr{C}$, we have

$$\widetilde{y} \cdot \mu = \widetilde{y'} \cdot \mu. \tag{16}$$

Since y and y' are nonterminal, they are reactant complexes. Hence, Eq. (16) may be written as

$$T_{\cdot,y} \cdot \mu = T_{\cdot,y'} \cdot \mu, \tag{17}$$

which is equivalent to Eq. (4) in Theorem 2.

The extension of the Shinar-Feinberg Theorem on ACR to PL-RDK systems is stated as follows.

Theorem 1. *Let $\mathscr{N} = (\mathscr{S}, \mathscr{C}, \mathscr{R})$ be a deficiency-one CRN and suppose that $(\mathscr{N}, K)$ is a PL-RDK system which admits a positive equilibrium. If $y, y' \in \mathscr{C}$ are nonterminal complexes whose kinetic order vectors differ only in species X_i, then the system has ACR in X_i.*

Proof. Suppose c^* and c^{**} are positive equilibria of the PL-RDK system $(\mathscr{N}, K)$. Observe that since $y, y' \in \mathscr{C}$ are nonterminal complexes whose kinetic order vectors differ only in species X_i, we have

$$T_{\cdot,y} - T_{\cdot,y'} = aX_i$$

for some nonzero a. Thus Eq. (4) reduces to

$$a(\log c_i^* - \log c_i^{**}) = 0.$$

It follows that

$$c_i^* = c_i^{**}.$$

That is, the system has ACR in species X_i.

The ODE system in Eq. (2) can be translated into a dynamically equivalent CRN with associated kinetic order matrix by employing the notion of **total CRN representation of Generalized Mass Action (GMA) systems**, proposed by Arceo et al. [5]. GMA system is a canonical framework used in Biochemical Systems Theory (BST) wherein every mass transfer rate is approximated separately with a power-law term, and these terms are added together, with a plus sign for incoming fluxes and a minus sign for outgoing fluxes [28,29]. For BST-related concepts, the reader may refer to the BST tutorial in the Appendix of Arceo et al. [3].

The total CRN representation of a GMA system allows for the CRN-based analysis of the dynamical system. Viewed as a GMA system, the set of ODEs in (2) has the following total CRN representation:

$$\begin{aligned} R_1 : \quad & X_2 \xrightarrow{k_1} X_1 \\ R_2 : X_1 + X_2 & \xrightarrow{k_2} 2X_2 \end{aligned} \tag{18}$$

with associated kinetic order matrix F given by

$$F = \begin{matrix} & \begin{matrix} X_1 & X_2 \end{matrix} \\ \begin{matrix} R_1 \\ R_2 \end{matrix} & \begin{bmatrix} 0 & 0.8 \\ 0.5 & 0.8 \end{bmatrix} \end{matrix}.$$

The CRN in (18) is a deficiency-one network with nonterminal complexes $X_1 + X_2$ and X_2 whose kinetic order rows differ only in X_1. The previous theorem indicates ACR in X_1, which agrees with the computation in (3).

The following simple proposition provides some examples for the ACR theorem for PL-RDK systems. As preparation, we recall some notions from [4,27] which are used in the result. A PL-RDK is said to be **reactant set linear independent** (of type **PL-RLK**) if the columns of T are linearly independent. We also recall the **reactant matrix** Y_{res}, which is obtained from the matrix representation of Y by removing the columns corresponding to non-reactant complexes. Its image Im Y_{res} is called the **reactant subspace** R, whose dimension q is called the **reactant rank** of the CRN. The **reactant deficiency** δ_ρ is the difference between the number of reactant complexes n_r and the reactant rank q.

Proposition 3. *Let $(\mathscr{S}, \mathscr{C}, \mathscr{R})$ be a deficiency one reaction network, which with PL-RDK, admits a positive equilibrium. Suppose the network has zero reactant deficiency, two nonterminal complexes $y, y' \in \mathscr{C}$ differing only in X_j and the map*

$$\widehat{y} := T \circ Y_{res}^{-1} : R \to Im\, T$$

is given by

$$\widehat{y}(X_1, \ldots, X_j, \ldots, X_m) = (a_1 X_1, \ldots, a_j X_j \ldots, a_m X_m), a_i \neq 0.$$

Then the system is PL-RLK and has ACR in X.

Proof. Since $\widehat{y}$ is an isomorphism, $T = \widehat{y} \circ Y_{\text{res}}$ is also an isomorphism. This implies that the system is PL-RLK. The kinetic order vector difference of y and y' is $(0, \ldots, ka_j, \ldots, 0)$ for some nonzero real k, so that Theorem 1's condition is fulfilled.

4 Application to a Carbon Cycle Model

The pre-industrial carbon cycle model of Anderies et al. [1] is a simple mass balance which involves three interacting carbon pools: land, atmosphere and ocean. Pictorially, the system can be depicted using a biochemical map comprised of nodes that represent carbon pools, solid arrows that indicate transfer of carbon, and dashed arrows that indicate if a pool affects or modulates a process. Figure 1 presents the biochemical map of the model of interest.

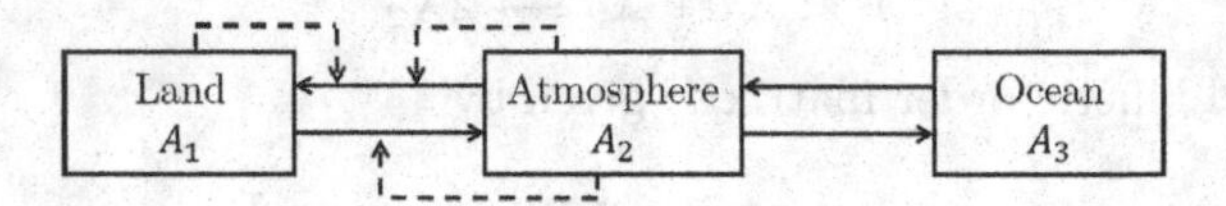

Fig. 1. Biochemical map of the pre-industrial carbon cycle model of Anderies et al. [1]

In our previous work [13], we reviewed the model's design and underlying assumptions and described the parameters and ODEs present in the pre-industrial state of the carbon cycle model. We also approximated all rate processes by products of power-law functions in order to obtain a GMA system approximation of the original system. The resulting ODEs of the approximation is given in (19):

$$\left.\begin{aligned} \dot{A}_1 &= k_1 A_1^{p_1} A_2^{q_1} - k_2 A_1^{p_2} A_2^{q_2} \\ \dot{A}_2 &= k_2 A_1^{p_2} A_2^{q_2} - k_1 A_1^{p_1} A_2^{q_1} - a_m A_2 + a_m \beta A_3 \\ \dot{A}_3 &= a_m A_2 - a_m \beta A_3, \end{aligned}\right\}. \tag{19}$$

We also obtained in [13], using total CRN representation of [5], the following deficiency-one CRN representation for the model:

$$\begin{aligned} A_1 + 2A_2 &\rightarrow 2A_1 + A_2 \\ A_1 + A_2 &\rightarrow 2A_2 \\ A_2 &\rightleftarrows A_3 \end{aligned} \tag{20}$$

Its associated kinetic order matrix is the transpose of the following T-matrix:

$$T = \begin{array}{c} \begin{array}{cccc} A_1 + 2A_2 & A_1 + A_2 & A_2 & A_3 \end{array} \\ \left[\begin{array}{cccc} p_1 & p_2 & 0 & 0 \\ q_1 & q_2 & 1 & 0 \\ 0 & 0 & 0 & 1 \end{array}\right] \begin{array}{c} A_1 \\ A_2 \\ A_3 \end{array} \end{array}. \tag{21}$$

In the Appendix, it is shown that there is a scenario in the pre-industrial state leading to a GMA system approximation, such that the kinetic order vectors of the nonterminal vertices A_1+2A_2 and A_1+A_2 differ only in A_2; that is, $p_1-p_2=0$ and $q_1-q_2 \neq 0$. In particular, this occurs when the human terrestrial carbon off-take term (which accounts for human activities that reduce the capacity of terrestrial pool to capture carbon such deforestation and land-use change) vanishes. Assuming the existence of a steady state, Theorem 1 indicates that the system has ACR in A_2. In fact, when $p_1=p_2$, steady state computation of the system in (19) yields the following equilibria set for the system:

$$E_+(\mathscr{N},K) = \left\{ \begin{bmatrix} A_1 \\ A_2 \\ A_3 \end{bmatrix} \in \mathbb{R}^{\mathscr{S}}_{>0} \;\middle|\; \begin{array}{l} A_2 = \left(\dfrac{k_2}{k_1}\right)^{\frac{1}{q_1-q_2}}, \\ A_3 = \dfrac{1}{\beta}\left(\dfrac{k_2}{k_1}\right)^{\frac{1}{q_1-q_2}}, \text{ and} \\ A_1 = A_0 - \left(1+\dfrac{1}{\beta}\right)\left(\dfrac{k_2}{k_1}\right)^{\frac{1}{q_1-q_2}} \end{array} \right\},$$

where A_0 = total conserved carbon at pre-industrial state.

5 Conclusion and Outlook

In conclusion, we summarize our results and outline some perspectives for further research.

1. We modified the Shinar-Feinberg Theorem on ACR for mass-action systems to include PL-RDK systems, a kinetic system more general than mass-action systems.
2. The theorem is applied to a power-law approximation of Anderies et al.'s Earth's carbon cycle in its pre-industrial state. The analysis reveals that there is a scenario in the pre-industrial state which yields a power-law approximation where there is ACR in the atmospheric carbon pool. Specifically, the power-law approximation leads to an ACR-possessing PL-RDK system when the human off-take coefficient, which accounts for the human activities that reduce the capacity of terrestrial pool to sequester carbon, vanishes.
3. The investigation of other forms of "concentration robustness" identified by Dexter et al. [10] for PL-RDK systems offers a further interesting research perspective.
4. The extension of the stochastic analysis of CRNs with ACR of Anderson et al. [2] for PL-RDK systems is another promising area for further investigation.

Acknowledgments. NTF acknowledges the support of the Department of Science and Technology-Science Education Institute (DOST-SEI), Philippines through the ASTHRDP Scholarship grant and Career Incentive Program (CIP). ARL and LFR held research fellowships from De La Salle University and would like to acknowledge the support of De La Salle University's Research Coordination Office.

Appendix

A Pre-industrial Carbon Cycle Model of Anderies et al.

The complete set of ODEs for the pre-industrial state is given by

$$\left.\begin{aligned} \dot{A}_1 &= r_{tc}[P(t) - R(t)]A_1\left[1 - \tfrac{A_1}{k}\right] - \alpha A_1 \\ \dot{A}_2 &= r_{tc}[R(t) - P(t)]A_1\left[1 - \tfrac{A_1}{k}\right] + \alpha A_1 - a_m A_2 + a_m \beta A_3 \\ \dot{A}_3 &= a_m A_2 - a_m \beta A_3. \end{aligned}\right\} \quad (22)$$

where

$$P(t) = a_f A_2(t)^{b_f} \cdot \left[a_p \cdot (a_T A_2(t) + b_T)^{b_p} \cdot e^{-c_p \cdot (a_T A_2(t) + b_T)}\right]$$

$$R(t) = \left[a_r \cdot (a_T A_2(t) + b_T)^{b_r} \cdot e^{-c_r \cdot (a_T A_2(t) + b_T)}\right].$$

For the description of the parameters, the reader is referred to [1] and the Appendix of [13]. The parameter values are identical to the values used in [13] but with $\alpha = 0$. This particular parameter is assigned to the human terrestrial carbon off-take rate. It is associated with human activities such as clearing, burning or farming, which reduce the capacity of land to capture carbon.

A power-law approximation of the ODE system at an operating point is obtained to generate a Generalized Mass Action (GMA) System [28,29]. Mathematically, GMA system approximation is equivalent to Taylor approximation up to the linear term in logarithmic space. The function $V(X_1, X_2, \ldots, X_m)$ can be approximated by $V = \alpha X_1^{p_1} X_2^{p_2} \cdots X_m^{p_m}$ at an operating point where

$$p_i = \frac{\partial V}{\partial X_i} \cdot \frac{X_i}{V} \text{ and } \alpha = V(X_1, X_2, \ldots, X_m) X_1^{-p_1} X_2^{-p_2} \cdots X_m^{-p_m}. \quad (23)$$

Table 1 presents the four carbon fluxes present in the pre-industrial state of the Anderies et al. model, and their corresponding rate functions. Furthermore, the last column lists their respective target power-law approximation. The last two functions, $a_m A_2$ and $a_m \beta A_3$, are already in the desired format and are thus, kept as is. To compute for the kinetic orders (and rate constants), we apply (23). By taking the parameter values used in [13] but with $\alpha = 0$, and assuming the initial values to be $A_1 = 2850/4500$, $A_2 = 750/4500$ and $A_3 = 900/4500$ (as in [1]), the ODE system in (22) reaches the following steady state: $A_1 = 0.7$, $A_2 = 0.15$ and $A_3 = 0.15$.

Table 1. Power-law approximation of the process rates.

Carbon flux	Function	Power-law approx.
$A_2 \rightarrow A_1$	$K_1 = r_{tc}P(t)A_1\left[1 - \frac{A_1}{k}\right]$	$k_1 A_1^{p_1} A_2^{q_1}$
$A_1 \rightarrow A_2$	$K_2 = r_{tc}R(t)A_1\left[1 - \frac{A_1}{k}\right] + \alpha A_1$	$k_2 A_1^{p_2} A_2^{q_2}$
$A_2 \rightarrow A_3$	$K_3 = a_m A_2$	$a_m A_2$
$A_3 \rightarrow A_2$	$K_4 = a_m \beta A_3$	$a_m \beta A_3$

```
G = af * A2^bf ;
T = aT * A2 + bT;
R = ar * T^br * Exp[-cr * T];
P = G * ap * T^bp * Exp[-cp * T] ;
H = a * A1;
K1 = rt * P * A1 * (1 - (A1 / k)) ;
K2 = rt * R * A1 * (1 - (A1 / k)) ;
K3 = am * A2;
K4 = am * B * A3;

p1 = Simplify[D[K1, A1] * (A1 / K1)]
q1 = Simplify[D[K1, A2] * (A2 / K1)]
k1 = K1 / ((A1^p1) * (A2^q1)) ;
```

$$\frac{2\,\mathrm{A1} - k}{\mathrm{A1} - k}$$

$$\frac{\mathrm{bf\,bT} - \mathrm{A2}^2\,\mathrm{aT}^2\,\mathrm{cp} + \mathrm{A2\,aT}\,(\mathrm{bf} + \mathrm{bp} - \mathrm{bT\,cp})}{\mathrm{A2\,aT} + \mathrm{bT}}$$

```
p2 = Simplify[D[K2, A1] * (A1 / K2)]
q2 = Simplify[D[K2, A2] * (A2 / K2)]
k2 = K2 / ((A1^p2) * (A2^q2)) ;
```

$$\frac{2\,\mathrm{A1} - k}{\mathrm{A1} - k}$$

$$-\frac{\mathrm{A2\,aT}\,(-\mathrm{br} + (\mathrm{A2\,aT} + \mathrm{bT})\,\mathrm{cr})}{\mathrm{A2\,aT} + \mathrm{bT}}$$

Fig. 2. Mathematica codes.

The algebraic calculations are implemented in Mathematica as shown in Fig. 2. When $\alpha = 0$ (i.e., the human off-take term vanishes),

$$p_1 = p_2 = \frac{2A_1 - k}{A_1 - k}.$$

For the power-law approximation, we choose values close to the equilibrium point as operating point: $A_1 = 0.69$, $A_2 = 0.155$ and $A_3 = 0.155$. Consequently, we obtain

$$\begin{aligned} p_1 &= -68, & p_2 &= -68, \\ q_1 &= 0.580148, & q_2 &= 0.910864. \end{aligned} \tag{24}$$

References

1. Anderies, J., Carpenter, S., Steffen, W., Rockström, J.: The topology of non-linear global carbon dynamics: from tipping points to planetary boundaries. Environ. Res. Lett. **8**(4), 044–048 (2013)

2. Anderson, D.F., Enciso, G.A., Johnston, M.D.: Stochastic analysis of biochemical reaction networks with absolute concentration robustness. J. R. Soc. Interface **11**(93), 20130943 (2014)
3. Arceo, C.P.P., Jose, E.C., Lao, A., Mendoza, E.R.: Reaction networks and kinetics of biochemical systems. Math. Biosci. **283**, 13–29 (2017)
4. Arceo, C.P.P., Jose, E.C., Lao, A.R., Mendoza, E.R.: Reactant subspaces and kinetics of chemical reaction networks. J. Math. Chem. **56**(2), 395–422 (2017). https://doi.org/10.1007/s10910-017-0809-x
5. Arceo, C.P.P., Jose, E.C., Marín-Sanguino, A., Mendoza, E.R.: Chemical reaction network approaches to biochemical systems theory. Math. Biosci. **269**, 135–52 (2015)
6. Bajzer, Z., Huzak, M., Neff, K.L., Prendergast, F.G.: Mathematical analysis of models for reaction kinetics in intracellular environments. Math. Biosci. **215**(1), 35–47 (2008)
7. Clarke, B.L.: Stoichiometric network analysis. Cell Biophys. **12**, 237–253 (1988)
8. Clegg, J.S.: Cellular infrastructure and metabolic organization. In: Stadtman, E.R., Chock, P.B. (eds.) From Metabolite, to Metabolism, to Metabolon. Current Topics in Cellular Regulation, vol. 33, pp. 3–14. Academic Press (1992)
9. Cortez, M.J., Nazareno, A., Mendoza, E.: A computational approach to linear conjugacy in a class of power law kinetic systems. J. Math. Chem. **56**(2), 336–357 (2017). https://doi.org/10.1007/s10910-017-0796-y
10. Dexter, J.P., Dasgupta, T., Gunawardena, J.: Invariants reveal multiple forms of robustness in bifunctional enzyme systems. Integr. Biol. **7**, 883–894 (2015)
11. Feinberg, M.: Lectures on chemical reaction networks. Notes of Lectures Given at the Mathematics Research Center of the University of Wisconsin (1979)
12. Feinberg, M.: The existence and uniqueness of steady states for a class of chemical reaction networks. Arch. Ration. Mech. Anal. **132**, 311–370 (1995)
13. Fortun, N., Lao, A., Razon, L., Mendoza, E.: A deficiency-one algorithm for a power-law kinetics with reactant determined interactions. J. Math. Chem. **56**(10), 2929–2962 (2018)
14. Horn, F., Jackson, R.: General mass action kinetics. Arch. Ration. Mech. Anal **47**, 187–194 (1972)
15. Kitano, H.: Biological robustness. Nat. Rev. Genet. **5**(11), 826–837 (2004)
16. Kopelman, R.: Rate processes on fractals: theory, simulations, and experiments. J. Stat. Phys. **42**, 185–200 (1986)
17. Kopelman, R.: Fractal reaction kinetics. Science **241**(4873), 1620–1626 (1988)
18. Kopelman, R., Koo, Y.: Reaction kinetics in restricted spaces. Isr. J. Chem. **31**(2), 147–157 (1991)
19. Kuthan, H.: Self-organisation and orderly processes by individual protein complexes in the bacterial cell. Prog. Biophys. Mol. Biol. **75**(1), 1–17 (2001)
20. Müller, S., Regensburger, G.: Generalized mass action systems: complex balancing equilibria and sign vectors of the stoichiometric and kinetic-order subspaces. J. SIAM Appl. Math. **72**, 1926–1947 (2012)
21. Müller, S., Regensburger, G.: Generalized mass-action systems and positive solutions of polynomial equations with real and symbolic exponents (invited talk). In: Proceedings of the International Workshop on Computer Algebra in Scientific Computing (CASC) (2014)
22. Newhouse, J.S., Kopelman, R.: Steady-state chemical kinetics on surface clusters and islands: segregation of reactants. J. Phys. Chem. **92**(6), 1538–1541 (1988)

23. Savageau, M.A.: Biochemical systems analysis: I. Some mathematical properties of the rate law for the component enzymatic reactions. Am. J. Sci. **25**(3), 365–369 (1969)
24. Savageau, M.A.: Development of fractal kinetic theory for enzyme-catalysed reactions and implications for the design of biochemical pathways. Biosystems **47**(1), 9–36 (1998)
25. Schnell, S., Turner, T.E.: Reaction kinetics in intracellular environments with macromolecular crowding: simulations and rate laws. Prog. Biophys. Mol. Biol. **85**(2–3), 235–260 (2004)
26. Shinar, G., Feinberg, M.: Structural sources of robustness in biochemical reaction networks. Science **327**(5971), 1389–1391 (2010)
27. Talabis, D.A.S.J., Arceo, C.P.P., Mendoza, E.R.: Positive equilibria of a class of power-law kinetics. J. Math. Chem. **56**(2), 358–394 (2017). https://doi.org/10.1007/s10910-017-0804-2
28. Voit, E.: Computational Analysis of Biochemical Systems: A Practical Guide for Biochemists and Molecular Biologists. Cambridge University Press, Cambridge (2000)
29. Voit, E.: Biochemical systems theory: a review. ISRN Biomath. **2013**, 1–53 (2013)
30. Wiuf, C., Feliu, E.: Power-law kinetics and determinant criteria for the preclusion of multistationarity in networks of interacting species. SIAM J. Appl. Dyn. Syst. **12**, 1685–1721 (2013)

Toward Unfolding Doubly Covered n-Stars

Hugo A. Akitaya[1], Brad Ballinger[2], Mirela Damian[3], Erik D. Demaine[4], Martin L. Demaine[4], Robin Flatland[5], Irina Kostitsyna[6], Jason S. Ku[10(✉)], Stefan Langerman[7], Joseph O'Rourke[8], and Ryuhei Uehara[9]

[1] University of Massachusetts, Lowell, USA
hugo_akitaya@uml.edu
[2] Humboldt State University, Arcata, USA
bradley.ballinger@humboldt.edu
[3] Villanova University, Villanova, USA
mirela.damian@villanova.edu
[4] Massachusetts Institute of Technology, Cambridge, USA
{edemaine,mdemaine}@mit.edu
[5] Siena College, Loudonville, USA
flatland@siena.edu
[6] Eindhoven University of Technology, Eindhoven, The Netherlands
i.kostitsyna@tue.nl
[7] Directeur de Recherches du F.R.S-FNRS, Université Libre de Bruxelles, Bruxelles, Belgium
stefan.langerman@ulb.ac.be
[8] Smith College, Northampton, MA, USA
jorourke@smith.edu
[9] Japan Advanced Institute of Science and Technology, Nomi, Japan
uehara@jaist.ac.jp
[10] National University of Singapore, Singapore, Singapore
jasonku@nus.edu.sg

Abstract. We present nonoverlapping general unfoldings of two infinite families of nonconvex polyhedra, or more specifically, zero-volume polyhedra formed by double-covering an n-pointed star polygon whose triangular points have base angle α. Specifically, we construct general unfoldings when $n \in \{3, 4, 5, 6, 8, 9, 10, 12\}$ (no matter the value of α), and we construct general unfoldings when $\alpha < 60°(1 + 1/n)$ (i.e., when the points are shorter than equilateral, no matter the value of n, or slightly larger than equilateral, especially when n is small). Whether all doubly covered star polygons, or more broadly arbitrary nonconvex polyhedra, have general unfoldings remains open.

Keywords: Polyhedra · Nonconvex · Nets · Nonoverlapping

1 Introduction

Unfolding a polyhedron P refers to the process of cutting and flattening its surface into a connected planar piece without overlap [13, Part III]. Finding

J. Akiyama et al. (Eds.): JCDCGGG 2018, LNCS 13034, pp. 122–135, 2021.
https://doi.org/10.1007/978-3-030-90048-9_10

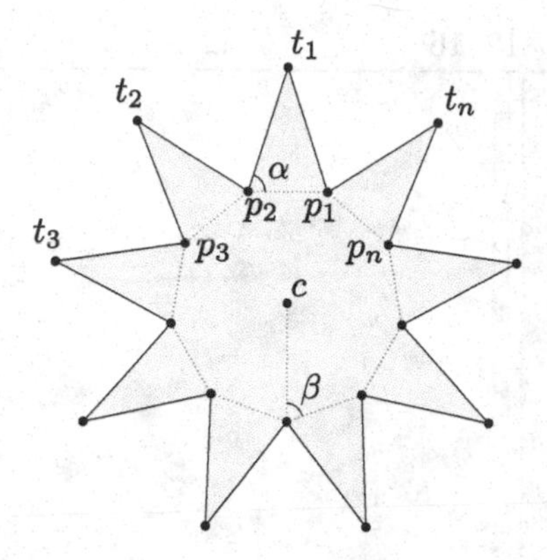

Fig. 1. An n-star polygon with base angle α, convex vertices (point tips) $t_1, t_2, \ldots, t_n$, and reflex vertices $p_1, p_2, \ldots, p_n$.

unfoldings is a classical problem with applications ranging from origami robots to sheet-metal manufacturing. Assuming P is genus 0, the cuts on the surface of P must form a forest to ensure a connected unfolding. To enable flattening into the plane, the cuts must span all (non-zero curvature) vertices of P, with at least one cut at each positive curvature vertex and two at each negative curvature vertex [13, Sec. 22.1.3]. A polyhedron admits many different unfoldings, depending on the choice of cuts.

Edge unfoldings restrict cuts to lie along the polyhedron's edges. A famous open problem (dating back to 1975 or even 1525) is whether every convex polyhedron has an edge unfolding. Recent progress solves this problem for "nearly flat convex caps" [19]. On the other hand, there are several examples of nonconvex polyhedra without edge unfoldings [3,4,14].

We focus in this paper on *general unfoldings*, which allow cuts anywhere on the polyhedron's surface. All convex polyhedra have general unfoldings by a variety of methods [2,12,16,20]. Thus we focus on general unfoldings of nonconvex polyhedra. There are nonconvex polyhedra with boundary, with no general unfoldings, but whether all nonconvex polyhedra without boundary have a general unfolding remains open [3].

The main progress on this problem has been for *orthogonal* polyhedra, whose edges and faces meet at right angles. All orthogonal polyhedra of genus ≤ 2 have general unfoldings [5–7,10]. For orthogonal polyhedra, we can quantify the simplicity of an unfolding by how close the cuts stick to the natural *grid* of the polyhedron, defined by extending planes through every face and taking all intersections with orthogonal faces. A *grid unfolding* sticks to cuts along this grid, while an $(a \times b)$-*grid unfolding* allows cuts on a *refined grid* defined by subdividing each grid face into an $a \times b$ subgrid for positive integers a, b. Grid unfoldings are known for *orthotubes* [4], well-separated *orthotrees* [9], and one-layer block structures [17], but it remains open whether they exist for all orthogonal polyhedra; see the survey [18]. The original method for unfolding all orthogonal polyhedra of genus 0 [10] uses exponential refinement. The level of refinement

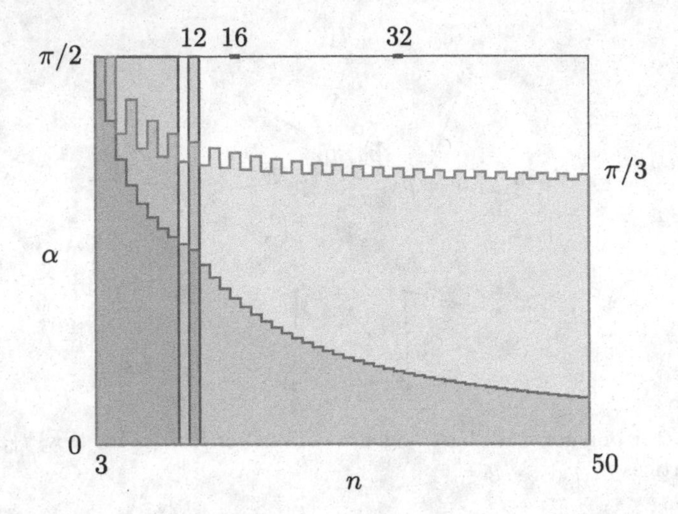

Fig. 2. Summary of (n, α)-stars whose unfoldings without overlaps are shown in this paper. Red is the naive unfolding, blue is the crown unfolding, purple shows comprehensive unfoldings. (Color figure online)

was later reduced to quadratic [6] and then linear [5], and finally generalized to genus-2 polyhedra (with linear refinement) [7]. Grid unfoldings that use sublinear refinement have been developed only for specialized orthogonal shape classes. For example, there exist (1×2)-grid unfoldings of orthostacks [4], (4×5)-grid unfoldings of Manhattan Towers [11], (2×1)-grid unfoldings of well-separated orthographs [15], and (4×4)-grid unfoldings of low-degree orthotrees [8].

Our Results. Apart from orthogonal polyhedra, we are not aware of any study of general unfoldings of infinite classes of nonconvex polyhedra. In this paper, we consider the seemingly simple class of *doubly covered polygons* formed by joining two copies of a simple polygon "back to back" by gluing corresponding edges, thereby forming a zero-volume genus-0 polyhedron. Specifically, we study *doubly covered* (n, α)*-stars*: for positive integer n and positive angle $\alpha < \pi/2$, an (n, α)-star is a simple polygon having $2n$ vertices $\{p_1, t_1, p_2, t_2, \ldots, p_n, t_n\}$ where points $p_1, p_2, \ldots, p_n$ form a regular n-gon, all points t_i lie outside of it, and for all $1 \leq i \leq n$ triangle $\Delta p_i t_i p_{i+1}$ is isosceles with $|p_i t_i| = |t_i p_{i+1}|$ and base angle $\angle p_i p_{i+1} t_i = \alpha$; see Fig. 1. (Throughout, we assume all vertex indices are computed modulo n.) We also let $\beta = \frac{\pi}{2}\left(1 - \frac{2}{n}\right)$ be half the angle of a regular n-gon. This polyhedron has exactly two faces, and does not admit an edge unfolding.

In this paper, we explore the space of doubly covered stars in search of families of general unfoldings, and search for counterexamples of polyhedra that do not admit a general unfolding. We show that general unfoldings of doubly covered (n, α)-stars exist:

- for any n when the base angle $\alpha < 60°(1 + 1/n)$, and
- for any base angle $\alpha \in (0, \pi/2)$ when $n \in \{3, 4, 5, 6, 7, 8, 9, 10, 12\}$.

These results are summarized in the plot in Fig. 2. We prove existence by construction, providing families of general unfoldings within specific subdomains of n and α.

There is a sense in which doubly covered polygons should be counted as "polyhedra" for the purposes of unfolding, because Alexandrov's gluing theorem, a fundamental tool in this area, includes these flat, zero-volume polyhedra [1]. Nevertheless, all our unfoldings are strictly non-overlapping, i.e., the unfoldings are simple polygons in the plane (as opposed to weakly simple polygons, which often arise when unfolding orthogonal polyhedra). Thus, there is a non-zero "clearance buffer" around the boundary of an unfolding that includes no points of the unfolding.

Now imagine inflating the flat, doubly covered P as follows. Surround the top side with a large radius r sphere S, with the spike tips of P touching S, and project the remaining vertices up to the surface of S, forming a positive-volume polyhedron P'. This alters the edge lengths and face angles by an amount that we can make arbitrarily small by choosing r sufficiently large. Thus we can select r so that the unfolding remains within the clearance buffer. So all our results also hold for these non-zero volume polyhedra P'.

2 Naive Unfolding

When the base angle α of a doubly covered (n, α)-star is small, we can produce a *naive unfolding* that is non-overlapping. For each i, we cut the top layer of the star along segments cp_i (center to spike base) and $p_i t_{i-1}$ (left side of spike). This cuts the top layer of the star into n quadrilaterals $Q_1, Q_2, \ldots, Q_n$, where $Q_i = \diamond p_i c p_{i+1} t_i$. The star can be unfolded along the right side $p_i t_i$ of each spike: each quadrilateral Q_i is reflected along the line $p_i t_i$. See Fig. 3 (left). We now prove:

Lemma 1. *The naive unfolding of an (n, α)-star is non-overlapping when*

$$\alpha \leq \begin{cases} \frac{\pi}{6}\left(1+\frac{6}{n}\right), & \textit{for } n \leq 12, \\ \frac{3\pi}{n}, & \textit{otherwise.} \end{cases}$$

Proof. We compute the values of n and α for which the naive unfolding is non-overlapping. Note that, due to symmetry, it is enough to show that one quadrilateral, say Q_1, unfolds without overlapping. Let c_i' and p_{i+1}' be the reflections of the points c and p_{i+1} across the line $p_i t_i$ respectively. The image of Q_i after unfolding is a quadrilateral $Q_i' = \diamond p_i c_i' p_{i+1}' t_i$. For the unfolding to be non-overlapping it is necessary that the total sum of the angles around point p_1 is not greater than 2π, that is $3\alpha + 3\beta \leq 2\pi$, which reduces to:

$$\alpha \leq \frac{\pi}{6}\left(1+\frac{6}{n}\right).$$

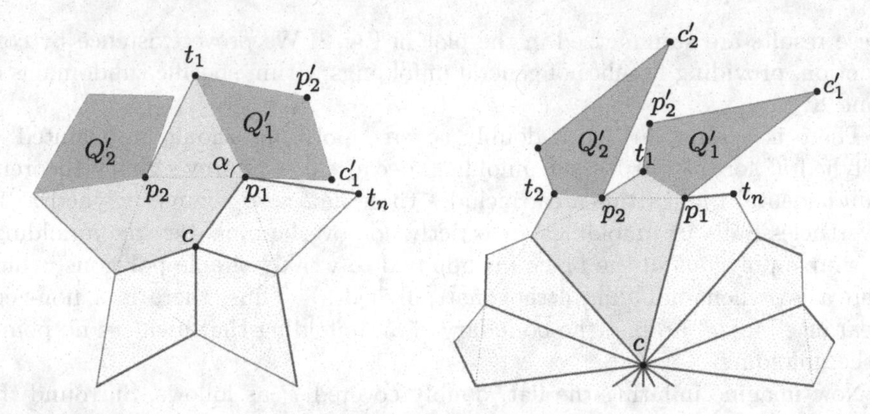

Fig. 3. Naive unfolding. The cuts and raw edges are shown in green; top layer pieces are shaded darker (blue) and bottom are lighter (yellow). Left: Two unfolded spikes for a star with $\alpha \geq \pi/4$. Right: two unfolded spikes for a star with $\alpha < \pi/4$. (Color figure online)

Assume that the above inequality holds, and note that Q'_1 lies on one side of the line p_1t_n. Now consider two cases, when the angle $\angle p_2t_1p'_2$ is non-reflex, and when it is reflex. In the first case, when $\angle p_2t_1p'_2 \leq \pi$, or equivalently $\alpha \geq \frac{\pi}{4}$, quadrilateral Q'_1 lies on one side of the line p_2t_1 (refer to Fig. 3). Therefore none of the images of the quadrilaterals Q_i overlap with each other, as they all lie inside the non-overlapping cones defined by the lines p_it_{i-1} and $p_{i+1}t_i$. That is, the unfolding is flat when

$$\frac{\pi}{4} \leq \alpha \leq \frac{\pi}{6}\left(1 + \frac{6}{n}\right).$$

This pair of inequalities has a solution when $0 < n \leq 12$.

In the second case, when $\angle p_2t_1p'_2 > \pi$, or equivalently $\alpha < \frac{\pi}{4}$, quadrilateral Q'_1 may overlap with Q'_2, the image of the quadrilateral Q_2, if the exterior angle $\angle t_1p_2c'_2$ is too small. Specifically, Q'_1 and Q'_2 will overlap if and only if p'_2 and t_1 lie on opposite sides of $p_2c'_2$, which happens when $\angle t_1p_2p'_2 = \frac{\pi-4\alpha}{2} > t_1p_2c'_2$. Thus, for the unfolding to be non-overlapping the total sum of the interior angles around p_2 has to be at most $2\pi - \frac{\pi-4\alpha}{2}$, which reduces to

$$\alpha \leq \frac{3\pi}{n}.$$

Putting the two cases together proves the lemma. □

3 Crown Unfolding

For doubly-covered (n, α)-stars with larger base angles, we can produce a non-overlapping *crown unfolding, so named for its four crown-like pieces shown in*

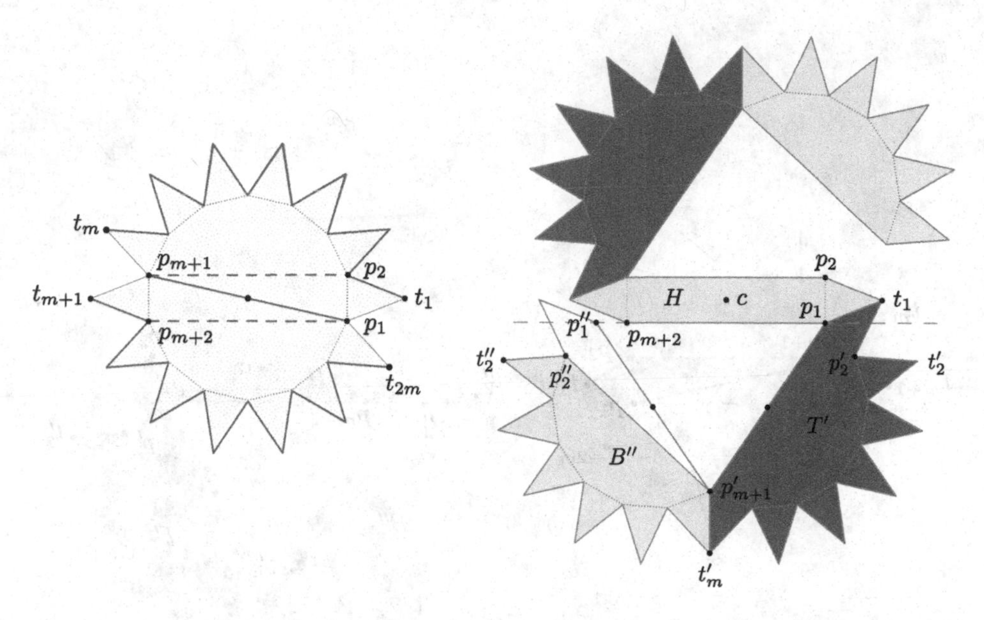

Fig. 4. Left: A doubly-covered $(2m, \alpha)$-star. Top layer cuts are shown as solid green lines and bottom layer cuts are dashed. Right: The resulting unfolding. Top layer pieces are shaded darker (blue) and bottom are lighter (yellow). The white quadrilateral is not a piece but is used to simplify the proof. (Color figure online)

Fig. 4(right). To produce this unfolding, first make one cut across the top layer of the star along segment p_1p_{m+1}, and two cuts across the bottom layer along p_2p_{m+1} and p_1p_{m+2}, where $m = \lceil n/2 \rceil$. Furthermore, cut along both edges of every spike, except for edges p_1t_1, p_1t_{2m}, $p_{m+1}t_m$, and $p_{m+1}t_{m+1}$; see Fig. 4 and Fig. 5 for even and odd n respectively. The top layer is cut into two crown-shaped pieces, while the bottom layer is cut into three pieces: a hexagon and two more crown-shaped pieces. The star can then be unfolded along the four preserved spike edges to form its crown unfolding. We now prove:

Theorem 1. *The crown unfolding of an (n, α)-star is non-overlapping when*

$$\alpha \leq \begin{cases} \frac{\pi}{3}\left(1 + \frac{2}{n}\right), & \textit{for even } n\,, \\ \frac{\pi}{3}\left(1 + \frac{1}{n}\right), & \textit{for odd } n\,. \end{cases}$$

Proof. We compute the values of n and α for which the crown unfolding is non-overlapping. Let T_1 and T_2 denote the two crown-shaped pieces from the top layer, where $T_1 = p_1t_1 \ldots t_mp_{m+1}$ and $T_2 = p_{m+1}t_{m+1} \ldots t_{2m}p_1$; let B_1 and B_2 denote the two crown-shaped pieces in the bottom layer, where $B_1 = p_2t_2 \ldots t_mp_{m+1}$ and $B_2 = p_{m+2}t_{m+2} \ldots t_{2m}p_1$; and let $H = p_1t_1p_2p_{m+1}$ $t_{m+1}p_{m+2}$ be the hexagonal bottom piece. We will consider the images of these pieces after unfolding, and will prove bounds on α for when they do not overlap.

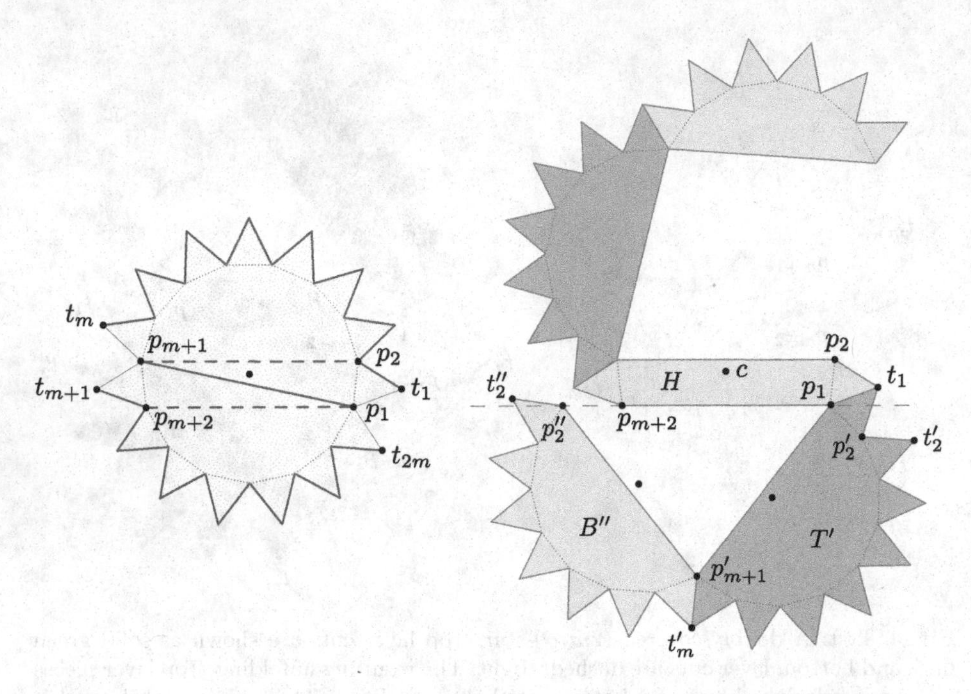

Fig. 5. Left: A doubly-covered $(2m+1, \alpha)$-star. Top layer cuts are shown as solid green lines and bottom layer cuts are dashed. Right: The resulting unfolding. Top layer pieces are shaded darker (blue) and bottom are lighter (yellow). (Color figure online)

Denote the reflections of the points p_i and t_i across edge p_1t_1 as p_i' and t_i' respectively, for all $1 \leq i \leq m+1$; and let $T' = p_1't_1' \ldots t_m'p_{m+1}'$ denote the image of the top layer piece T_1 after unfolding across edge p_1t_1. Then, let p_i'' and t_i'' be the reflections of the points p_i' and t_i' across $t_m'p_{m+1}'$, for all $1 \leq i \leq m+1$. The bottom layer piece B_1 (that was attached to the uncut spike edge t_mp_{m+1}) then unfolds into the polygon $B'' = p_2''t_2'' \ldots t_m''p_{m+1}''$.

Consider the images T' of T_1 and B'' of B_1, which unfold on the same side relative to the hexagonal bottom piece H. Polygons H and T' do not overlap as they share edge p_1t_1 which lies on the convex hulls of both polygons. Similarly, polygons T' and B'' do not overlap because they share edge $t_m'p_{m+1}'$. It is not hard to see that the crown unfolding does not intersect for sufficiently small α (say, $\alpha < \frac{\pi}{4}$) as T' and B'' exist on one side of the extension of p_1t_1, with the rest of the unfolding on the other side. Thus we focus our attention on the upper bound, for $\alpha \geq \frac{\pi}{4}$.

First we consider the case when n is even and prove that the crown unfolding is non-overlapping when $\alpha \leq \alpha^*_{even}$ where $\alpha^*_{even} = \frac{\pi}{3}\left(1 + \frac{2}{n}\right)$; see Fig. 4. Observe that the reflection of T' across $t_m'p_{m+1}'$ contains B'' and point p_1''. Then, when p_1'' is in the half-plane Π bounded by p_1 and p_{m+2} not containing p_2, then T' and B'' will both exist in the half-plane Π' bounded by t_1 and t_{m+1} and will not overlap with the rest of the unfolding. Define angles:

$$\theta_{even} = \angle p_{m+2}p_1p'_{m+1} = \frac{3\pi}{2} - 2\alpha - \beta \text{ and } \phi_{even} = \angle p_1p'_{m+1}p''_1 = 2\pi - 2\alpha - 2\beta.$$

Observe that p''_1 is in Π when $\phi_{even} + 2\theta_{even} \geq \pi$, which holds exactly when $\alpha \leq \alpha^*_{even}$; so T' and B'' both stay in P' when $\alpha \leq \alpha^*_{even}$. Further, define distances $d_1 = |p_1p_{m+2}|$, $d_2 = |p_1p'_{m+1}|$, and $d_3 = |p_1p''_1|$. Clearly $d_1 < d_2$ since d_2 is a diameter of the n-gon and d_1 is a shorter diagonal, and when $\alpha \leq \alpha^*_{even}$, $d_2 < d_3$, since $\phi_{even} = \pi + \frac{2\pi}{n} - 2\alpha \geq \frac{\pi}{3} + \frac{2\pi}{3n} > \frac{\pi}{3}$. Therefore, $d_1 < d_3$, and B'' also does not intersect H.

By symmetry, the same argument also shows that the images of T_2 and B_2 do not overlap with H after unfolding, and lie on the other side of the line through t_1 and t_{m+1}. Thus, the crown unfolding of a doubly-covered (n, α)-star is non-overlapping when n is even and $\alpha \leq \alpha^*_{even}$.

Now we consider the case when n is odd and prove that the crown unfolding is non-overlapping when $\alpha \leq \alpha^*_{odd}$ where $\alpha^*_{odd} = \frac{\pi}{3}\left(1 + \frac{1}{n}\right)$; see Fig. 5. First we observe that, when α increases, the unfolded images T' and B'' of T_1 and B_1 approach H faster than the unfolded images of T_2 and B_2, so we focus our attention on T' and B''. We first observe that distance $|p_1p'_{m+1}|$ is the same as $|p'_{m+1}p''_2|$. Then, when p''_2 is in the half-plane Π bounded by p_1 and p_{m+2} not containing p_2, then T' and B'' will both exist in the half-plane Π' bounded by t_1 and t_{m+1} and will not overlap with the rest of the unfolding. Define angles:

$$\theta_{odd} = \angle p_{m+2}p_1p'_{m+1} = \frac{3\pi}{2} - 2\alpha - \beta - \frac{\pi}{n} \text{ and } \phi_{odd} = \angle p_1p'_{m+1}p''_1 = 2\pi - 2\alpha - 2\beta.$$

Observe that p''_2 is in P when $\phi_{odd} + 2\theta_{odd} \geq \pi$, which holds exactly when $\alpha \leq \alpha^*_{odd}$; so T' and B'' both stay in P' when $\alpha \leq \alpha^*_{odd}$. Further, define distances $d_1 = |p_1p_{m+2}|$, $d_2 = |p_1p'_{m+1}|$, and $d_3 = |p_1p''_2|$. Clearly $d_1 < d_2$ since d_2 is a longer diagonal of the n-gon than d_1. When $\alpha \leq \alpha^*_{odd}$, $d_2 < d_3$ since $\phi_{odd} = \pi + \frac{2\pi}{n} - 2\alpha \geq \frac{\pi}{3} + \frac{4\pi}{3n} > \frac{\pi}{3}$; so $d_1 < d_3$, and B'' does not intersect H. Thus, the crown unfolding of a doubly-covered (n, α)-star is also non-overlapping when n is odd and $\alpha \leq \alpha^*_{odd}$, completing the proof. □

4 Comprehensive Unfoldings

In this section, we provide unfoldings of (n, α)-stars for arbitrary α for some small choices of n. We call such unfoldings *comprehensive unfoldings.*

Theorem 2. *There exist non-overlapping unfoldings of (n, α)-stars for any positive $\alpha < \pi/2$ and any $n \in \{3, 4, 5, 6, 7, 8, 9, 10, 12\}$.*

Proof (Case $n = 3$).
The naive unfolding suffices, since $\alpha \leq \frac{\pi}{6}\left(1 + \frac{6}{3}\right) = \frac{\pi}{2}$. □

Proof (Case $n = 4$).
The crown unfolding suffices, since $\alpha \leq \frac{\pi}{3}\left(1 + \frac{2}{4}\right) = \frac{\pi}{2}$. □

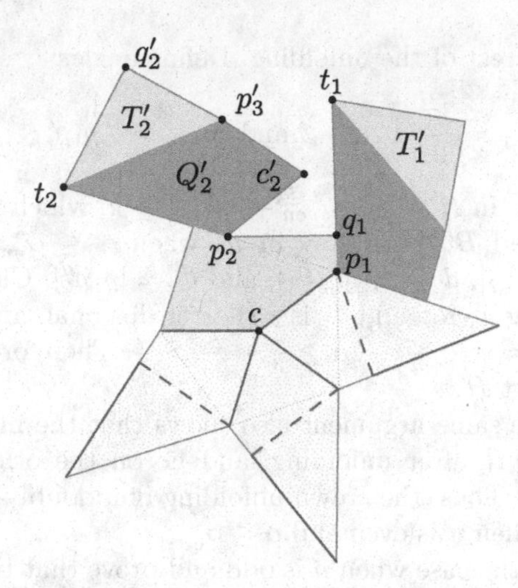

Fig. 6. General unfolding for $n = 5$. Top layer cuts are shown as solid green lines and bottom layer cuts are dashed. Top layer pieces are shaded darker (blue) and bottom are lighter (yellow). Only the first and second spikes are shown unfolded. Note that c_1' is under the fifth spike, but it will be uncovered when the fifth spike is unfolded. (Color figure online)

Proof (Case $n = 5$). The naive unfolding suffices for small $\alpha \le \frac{\pi}{3} < \frac{\pi}{6}\left(1 + \frac{6}{5}\right)$. It fails for larger α because the unfolded image of a top layer's spike intersects one of the bottom layer's spikes. Here we adopt an approach similar to the naive unfolding, but resolve the overlap by moving spike material from one side to the other; see Fig. 6.

To produce a comprehensive unfolding of a $(5, \alpha)$-star, first cut the top layer of the star along segments cp_i for all i; and also along the right side of the spike from t_i to q_i and along the bottom face to p_{i+1}, where q_i is the point on $p_i t_i$ such that $\angle p_i p_{i+1} q_i = \frac{\pi}{10}$. The bottom layer is cut into a central 10-gon B and five triangles T_i where $B = p_1 q_1 p_2 \ldots p_n q_n$ and $T_i = \Delta t_i p_{i+1} q_i$; while the top layer is cut into five quadrilaterals $Q_i = \diamond p_i c p_{i+1} t_i$. Note that Q_i remains attached to B alongside $p_i q_i$, and T_i remains attached to Q_i alongside $t_i p_{i+1}$.

Due to symmetry, it is sufficient to show that one spike unfolds without overlap. Let c_i' and p_{i+1}' be the reflections of c and p_{i+1} across line $p_i t_i$ respectively, and let $Q_i' = \diamond p_i c_i' p_{i+1}' t_i$ be the reflection of Q_i across line $p_i t_i$. Further, let q_i' be the reflection of q_i across line $p_{i+1}' t_i$, and let $T_i' = t_i p_{i+1}' q_i'$. It suffices to show that the total sum of angles around point p_i is not greater than 2π and that $p_i c_i'$ does not intersect $q_{i-1} t_{i-1}$ for any $\frac{\pi}{3} \le \alpha \le \frac{\pi}{2}$. The first condition is satisfied by construction, as $2\alpha + 3\beta + \frac{\pi}{10} \le 2\pi$ implies $\alpha \le \frac{\pi}{2}$. To prove the second condition, the closest any point on $q_{i-1} t_{i-1}$ gets to p_i, for $\frac{\pi}{3} \le \alpha \le \frac{\pi}{2}$, is $|p_i p_{i+1}| \sin \alpha \ge |p_i p_{i+1}| \sin(\frac{\pi}{3}) = |p_i p_{i+1}| \frac{\sqrt{3}}{2} \approx 0.866 |p_i p_{i+1}|$. Simple calculations show that the distance from c_i' to p_i is $|p_i c_i'| = \frac{1}{2 \cos \beta} |p_i p_{i+1}| = \sqrt{\frac{2}{5 - \sqrt{5}}} |p_i p_{i+1}| \approx$

$0.851|p_i p_{i+1}|$. These together show that p_i and c'_i lie on the same side of $q_{i-1}t_{i-1}$, so the unfolding does not overlap. □

Proof (Case $n = 6$). The naive unfolding suffices for small $\alpha \leq \frac{\pi}{6}\left(1 + \frac{6}{6}\right) = \frac{\pi}{3}$, so here we focus on larger angles $\alpha > \frac{\pi}{3}$. We cannot solve $n = 6$ in the same way as $n = 5$ because the center point c'_i would approach p_{i-1} as α approaches $\frac{\pi}{2}$, and would cut off the next spike. We solve this problem by unfolding two spikes at once; see the left of Fig. 7. While the $n \in \{3, 5\}$ comprehensive unfoldings each had n-fold rotational symmetry, the $n \in \{6, 9\}$ comprehensive unfolding constructions we present each have 3-fold rotational symmetry, and the $n \in \{8, 12\}$ each have 4-fold rotational symmetry.

To produce a comprehensive unfolding of a $(6, \alpha)$-star, first cut the top layer of the star along segments cp_i, but only for even i; and also along the right side of every odd spike, from t_i to p_i for odd i, and along the left side of every even spike, from t_i to p_{i+1} for even i. The bottom layer is cut into a central 9-gon B and three triangles T_i, where $B = p_1p_2t_2p_3p_4t_4p_5p_6t_1$ and $T_i = \triangle t_ip_{i+1}p_i$ for odd i; while the top layer is cut into three hexagons $H_i = \bigcirc t_ip_icp_{i+2}t_{i+1}p_{i+1}$ for even i. Note that H_i remains attached to B alongside t_ip_i, and T_{i+1} remains attached to H_i alongside $t_{i+1}p_{i+2}$.

Due to symmetry, it is sufficient to show that one spike unfolds without overlap. Let c'_i, p'_{i+1}, t'_{i+1}, and p'_{i+2} be the reflections of c, p_{i+1}, t_{i+1}, and p_{i+2} across line p_it_i respectively for even i, and let $H'_i = \bigcirc t_ip_ic'_ip'_{i+2}t'_{i+1}p'_{i+1}$ be the reflection of H_i across line p_it_i. Further, let p''_i be the reflection of p'_i across line $p'_{i+1}t'_i$ for odd i, and let $T''_i = t'_ip'_{i+1}p''_i$ also for odd i. It suffices to show that the total sum of angles around point p_i for even i is not greater than 2π, and that points c'_i and p''_{i+1} lie on the counterclockwise side of $p_{i-1}t_{i-2}$ for even i, for any $\frac{\pi}{3} \leq \alpha \leq \frac{\pi}{2}$. The first condition is satisfied by construction, as $2\alpha + 3\beta \leq 2\pi$ implies $\alpha \leq \frac{\pi}{2}$. To prove the second condition, we observe that points c'_i and p''_{i+1} always stay on the counterclockwise side of the ray R extending from p_{i-1} away from the center in the direction perpendicular to $p_{i-2}p_{i-1}$, while t_{i-2} always stays on the clockwise side of R, for any $\frac{\pi}{3} \leq \alpha \leq \frac{\pi}{2}$. We have found a dividing line separating the leaves of the unfolding, thus the unfolding does not overlap. □

For the remaining cases, we omit full details of the construction, and instead provide a sketch of how they were discovered and adapted.

Proof (Sketch for case $n = 9$).

Just as with $n = 6$, for $n = 9$, we target 3-fold symmetry, so as to allow the top layer center point to move farther away than the naive unfolding would achieve. The center n-gon of the top face is divided into three equal pieces and, as with the $n = 5$ comprehensive unfolding, additional cuts are made and propagated to avoid overlap for larger α; see the right side of Fig. 7. The proof of correctness is tedious, but follows the same proof techniques as used earlier in this paper, ensuring that the unfolding of spike i (and everything attached to it) never intersects the counterclockwise side of spike $i - 1$. □

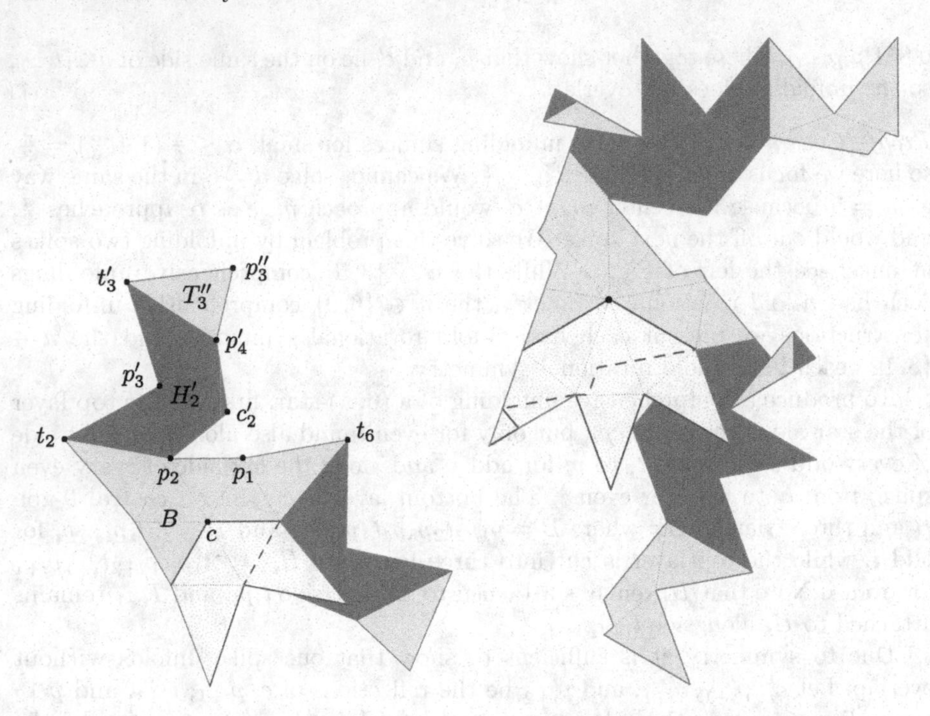

Fig. 7. General unfoldings for $n = 6$ [Left] and $n = 9$ [Right] . Top layer cuts are shown as solid green lines and bottom layer cuts are dashed. Top layer pieces are shaded darker (blue) and bottom are lighter (yellow). (Color figure online)

Fig. 8. General unfoldings for $n = 8$ [Left] and $n = 12$ [Right]. Top layer cuts are shown as solid green lines and bottom layer cuts are dashed. Top layer pieces are shaded darker (blue) and bottom are lighter (yellow). (Color figure online)

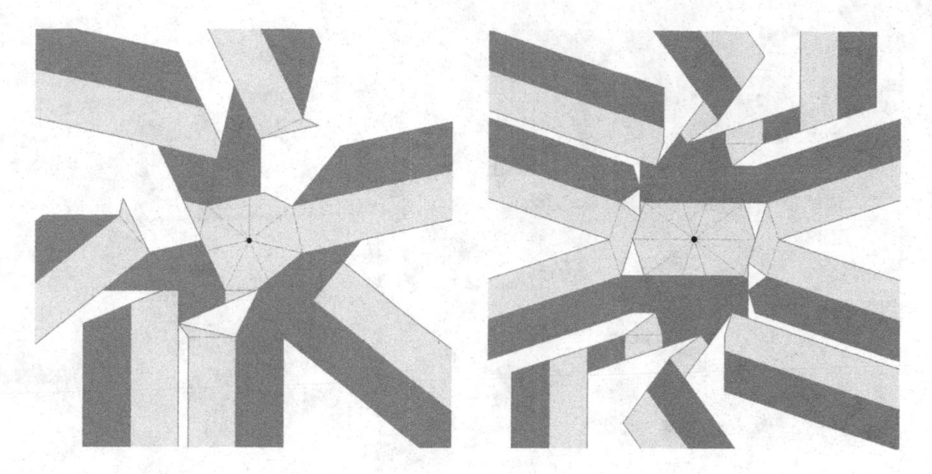

Fig. 9. General unfoldings for $n = 7$ [Left] and $n = 10$ [Right]. The cuts and raw edges are shown in green; top layer pieces are shaded darker (blue) and bottom are lighter (yellow). (Color figure online)

Proof (Sketch for case $n = 8$).

For $n = 8$, we choose to divide the top face, this time with 4-fold rotational symmetry; see the left of Fig. 8. The authors also have another comprehensive unfolding construction for $n = 8$ based on 2-fold rotational symmetry, but have found that 4-fold approaches tend to be easier to implement than 2-fold approaches. As before, additional cuts are made to avoid overlap for larger α. □

Proof (Sketch for case $n = 12$).

As with $n = 8$, we provide a comprehensive unfolding for $n = 12$ that has 4-fold symmetry, shown on the right of Fig. 8. We also have a comprehensive unfolding for $n = 12$ that has 2-fold symmetry, but it is substantially more complex. The figure shows an unfolding for an α angle close to $\frac{\pi}{2}$. □

Proof (Sketch for case $n = 7$).

Unlike $n = 5$, for which we could adapt the naive unfolding fairly directly, $n = 7$ is the first prime number for which the naive unfolding maps c'_i to a point in the interior of the original n-gon bottom layer face. Thus, we attempt to generalize the 3-fold symmetry approach of $n \in \{6, 9\}$ to unfold $n = 7$ in three pieces, two of which unfold two sectors of the original 7-gon while the third piece unfolds three sectors; see left of Fig. 9. □

Proof (Sketch for case $n = 10$).

While not prime, we were unable to find a comprehensive unfolding for $n = 10$ having 5-fold symmetry. Instead, we attempted to generalize the crown unfolding having 2-fold rotational symmetry, cutting away material that would cause the unfolding to intersect; see right of Fig. 9. Note that, while the crown unfolding retains exactly two whole sectors of the original n-gon from the bottom layer at

Fig. 10. Possible unfoldings for $n = 16$ [Left] and $n = 32$ [Right] for α close to $\frac{\pi}{2}$.

the center of the unfolding, this unfolding retains four sectors in the center. We have also used this construction approach as the basis for various 2-fold rotational symmetry comprehensive unfoldings, specifically for $n \in \{6, 8, 10, 12\}$. □

5 Future Work

Our initial goal in studying doubly covered stars was to find a counterexample to the proposition that every polyhedron admits a general unfolding. While we have not yet found a counterexample, our search resulted in a number of interesting unfoldings for a new generalized family of nonconvex polyhedra. Our exploration suggests that (n, α)-stars with large n and α close to $\frac{\pi}{2}$ may be potential counterexample candidates. We have attempted non-overlapping unfoldings for n larger than 12, for example for $n \in \{16, 32\}$ for α close to $\frac{\pi}{2}$ (as shown in Fig. 10), but we are not at all convinced that these unfoldings can be generalized for arbitrary α. We hope that future exploration of doubly-covered star unfoldings will either provide unfoldings for the remaining space over n and α, or find a polyhedron that does not admit a general unfolding.

References

1. Alexandrov, A.D.: Convex Polyhedra. Springer-Verlag, Berlin (2005). monographs in Mathematics. Translation of the 1950 Russian edition by N. S. Dairbekov, S. S. Kutateladze, and A. B. Sossinsky
2. Aronov, B., O'Rourke, J.: Nonoverlap of the star unfolding. Discret. Comput. Geom. **8**(3), 219–250 (1992). https://doi.org/10.1007/BF02293047
3. Bern, M., Demaine, E., Eppstein, D., Kuo, E., Mantler, A., Snoeyink, J.: Ununfoldable polyhedra with convex faces. Comput. Geom. Theory Appl. **24**(2), 51–62 (2003)

4. Biedl, T., et al.: Unfolding some classes of orthogonal polyhedra. In: Proceedings of the 10th Canadian Conference on Computational Geometry. Montréal, Canada (August 1998). http://cgm.cs.mcgill.ca/cccg98/proceedings/cccg98-biedl-unfolding.ps.gz
5. Chang, Y.-J., Yen, H.-C.: Unfolding orthogonal polyhedra with linear refinement. In: Elbassioni, K., Makino, K. (eds.) ISAAC 2015. LNCS, vol. 9472, pp. 415–425. Springer, Heidelberg (2015). https://doi.org/10.1007/978-3-662-48971-0_36
6. Damian, M., Demaine, E., Flatland, R.: Unfolding orthogonal polyhedra with quadratic refinement: the Delta-unfolding algorithm. Graphs and Combinatorics **30**(1), 125–140 (2014)
7. Damian, M., Demaine, E., Flatland, R., O'Rourke, J.: Unfolding genus-2 orthogonal polyhedra with linear refinement. Graphs Comb. **33**(5), 1357–1379 (2017)
8. Damian, M., Flatland, R.: Unfolding low-degree orthotrees with constant refinement. In: The 30th Canadian Conference on Computational Geometr (CCCG) 2018, pp. 189–208 (2018)
9. Damian, M., Flatland, R., Meijer, H., O'Rourke, J.: Unfolding well-separated orthotrees. In: The 15th Annual Fall Workshop on Computational Geometry and Visualization (FWCG) 2005, pp. 25–26 (2005)
10. Damian, M., Flatland, R., O'Rourke, J.: Epsilon-unfolding orthogonal polyhedra. Graphs Comb. **23**(1), 179–194 (2007)
11. Damian, M., Flatland, R., O'Rourke, J.: Unfolding Manhattan towers. Comput. Geom. Theory Appl. **40**, 102–114 (2008)
12. Demaine, E.D., Lubiw, A.: A generalization of the source unfolding of convex polyhedra. In: Márquez, A., Ramos, P., Urrutia, J. (eds.) EGC 2011. LNCS, vol. 7579, pp. 185–199. Springer, Heidelberg (2012). https://doi.org/10.1007/978-3-642-34191-5_18
13. Demaine, E.D., O'Rourke, J.: Geometric Folding Algorithms: Linkages, Origami, Polyhedra. Cambridge University Press, Cambridge (July 2007). http://www.gfalop.org
14. Grünbaum, B.: No-net polyhedra. Geombinatorics **11**, 111–114 (2002). http://www.math.washington.edu/~grunbaum/Nonetpolyhedra.pdf
15. Ho, K.-Y., Chang, Y.-J., Yen, H.-C.: Unfolding some classes of orthogonal polyhedra of arbitrary genus. In: Cao, Y., Chen, J. (eds.) COCOON 2017. LNCS, vol. 10392, pp. 275–286. Springer, Cham (2017). https://doi.org/10.1007/978-3-319-62389-4_23
16. Itoh, J.I., O'Rourke, J., Vîlcu, C.: Star unfolding convex polyhedra via quasigeodesic loops. Discret. Comput. Geom. **44**(1), 35–54 (2010)
17. Liou, M.-H., Poon, S.-H., Wei, Y.-J.: On edge-unfolding one-layer lattice polyhedra with cubic holes. In: Cai, Z., Zelikovsky, A., Bourgeois, A. (eds.) COCOON 2014. LNCS, vol. 8591, pp. 251–262. Springer, Cham (2014). https://doi.org/10.1007/978-3-319-08783-2_22
18. O'Rourke, J.: Unfolding orthogonal polyhedra. In: Surveys on Discrete and Computational Geometry: Twenty Years Later, Contemporary Mathematics, vol. 453, pp. 307–317. American Mathematical Society (2008). https://doi.org/10.1090/conm/453
19. O'Rourke, J.: Edge-unfolding nearly flat convex caps. CoRR abs/1707.01006 (2017). http://arxiv.org/abs/1707.01006
20. Sharir, M., Schorr, A.: On shortest paths in polyhedral spaces. SIAM J. Comput. **15**(1), 193–215 (1986)

Crystallographic Flat Origami from n-Uniform Tilings

Ma. Louise Antonette N. De Las Peñas[1(✉)] and Eduard C. Taganap[2]

[1] Department of Mathematics, Ateneo de Manila University, Quezon City, Philippines
mdelaspenas@ateneo.edu

[2] Department of Mathematics and Physics, Central Luzon State University, Science City of Muñoz, Philippines

Abstract. This paper discusses the symmetry properties of crystallographic flat origami arising from n-uniform tilings using the hinged tiling method. A flat origami invariant under a plane crystallographic group is called a *crystallographic flat origami*. An *n-uniform tiling* is a tiling consisting of regular polygons, with the property that its vertices form n transitivity classes under the action of its symmetry group.

Keywords: Crystallographic flat origami · Flat origami · n-uniform tiling · Origami symmetry

1 Introduction

In the past years, the centuries old art of paper folding or *origami* has evolved into an area of study called origami science and mathematics, and is fast expanding to include origami in education, design, and origami applications in engineering and technology [21]. A problem on origami that continues to be a focus of attention in mathematical studies pertains to flat foldability (whether a crease pattern can be folded into a flat origami - folded state that lies in the plane) and whether such a flat origami becomes crystallographic in nature. If a flat origami is invariant under a plane crystallographic group, we call it a *crystallographic flat origami* or an *origami tessellation*.

The term crystallographic flat origami was defined in [11], where mathematical properties of this class of flat origami were first discussed. Since then, there have been several works on crystallographic flat origami. In [20], Verill discussed different ways of constructing flat foldable crease patterns derived from Archimedean tilings, which may give rise to crystallographic flat origami. Lang, in his published book [14], presented various ways of arriving at flat foldable crease patterns derived from different classes of tilings, which folds to crystallographic flat origami. In [1], Bateman presented a computer program named *Tess*,

The authors would like to thank the Ateneo de Manila University for the support through the Loyola Schools Scholarly Work Grant.

J. Akiyama et al. (Eds.): JCDCGGG 2018, LNCS 13034, pp. 136–151, 2021.
https://doi.org/10.1007/978-3-030-90048-9_11

that can be used in designing unassigned crease patterns for crystallographic flat origami. In [3,17,19], the use of color symmetry in arriving at crystallographic flat origami obtained from Archimedean tilings was discussed.

An *n-uniform tiling* is a tiling by regular polygons whose vertices form n transitivity classes under the action of its symmetry group. A 1-uniform tiling is called an *Archimedean tiling*.

This paper affirms that a crystallographic flat origami can be constructed from an n-uniform tiling; and applying the hinged tiling method to the tiling yields a locally flat foldable crease pattern X that results in a crystallographic flat origami invariant under direct isometries. A systematic construction of possible crystallographic origami resulting from X will be given in this paper. The method provides a way of arriving at the different MV-assignments to a locally flat foldable crease pattern arising from a tiling that has more than one transitivity class of vertices. Moreover, the method which is based on group theoretic notions, facilitates the characterization of the symmetry group types of the resulting crystallographic origami. This paper also gives a result on *iso-area crystallographic flat origami* - origami where equal portions of each side of the paper are shown on the exterior of the origami [12].

2 Flat Origami

A *flat origami* is a folded single piece of paper that can be pressed in a book without crumpling. A way to study flat origami is through its crease pattern. A *crease pattern* is a collection of segments, called *crease lines* indicating where the folds are to be made and how the paper will be folded. The crease lines may be folded either as a mountain (M) or valley (V) fold. When the paper is folded, mountain folds form protruding ridges while valley folds form indented troughs. We call the M and V fold assignments to the crease lines of the crease pattern an *MV-assignment*.

A crease pattern is said to be *flat foldable* if it is a crease pattern of some flat origami. Flat foldability of a crease pattern is a focus of interest in mathematical origami research. The flat foldability of a single vertex crease pattern has been well-studied by Kawasaki [9], Maekawa (see [13]), Justin [8] and Hull [6,7]. A necessary and sufficient condition for a single-vertex crease pattern to be folded flat is determined by both Kawasaki and Justin independently, which is given in the following theorem.

Theorem 1 [6,8,9]. *A single-vertex crease pattern, where the angles between the crease lines are given by* $\theta_1, \theta_2, \ldots, \theta_n$ *(in this order), is flat foldable if and only if* n *is even and the sum of alternate angles is equal to* π *or equivalently* $\theta_1 + \theta_3 + \cdots + \theta_{n-1} = \theta_2 + \theta_4 + \cdots + \theta_n = \pi$.

In general, determining whether a multi-vertex crease pattern is flat foldable is an NP-hard problem [2]. Thus, we only consider the crease patterns that are locally flat foldable and fold them to check whether these actually fold flat. By a *locally flat foldable crease pattern* we mean a crease pattern that satisfies the Kawasaki's angle criterion at every vertex of the crease pattern.

A locally flat foldable crease pattern can be obtained in several ways. In this text, we take advantage of a method called the *hinged tiling* method. The hinged tiling method was developed by Palmer, Barreto and Bateman and was applied to an Archimedean tiling to arrive at a locally flat foldable crease pattern in [20]. In [15], Lang and Bateman also discussed the hinged tiling method, which they called the *shrink-rotate algorithm*, and showed that flat origami can be created from n-uniform tilings. In this paper, we apply the hinged tiling method to n-uniform tilings. This result in a wider variety of locally flat foldable crease patterns and will be given suitable MV-assignments that give rise to particular classes of crystallographic flat origami.

The steps in the hinged tiling method are:

(a) Consider an n-uniform tiling (its tiles called *primal*) and its dual tiling (its tiles called *dual* tiles).
(b) Shrink the tiles in the n-uniform tiling.
(c) Insert the dual tiles between the primal tiles.
(d) Rotate the primal tiles.
(e) Scale the dual tiles relative to the primal tiles as desired.

The resulting crease pattern is a combination of regular polygon twists. A *regular polygon twist* is a crease pattern consisting of a central regular polygon surrounded by *pleats*, which are pairs of parallel crease lines emanating from the endpoints of a single side of the central polygon. The *twist angle* ρ of a regular polygon twist is the angle formed by a side of the central polygon and a crease line of a pleat. In general, for a regular k-gon twist to avoid paper overlapping or crossing when folded, we impose the condition that $0 < \min(\rho, \pi - \rho) < \frac{\pi}{2} - \frac{\pi}{k}$ [14]. So in step (d), we impose the condition that the angle of rotation ρ is such that $0 < \rho < \frac{\pi}{2} - \frac{\pi}{m}$, where an m-gon is the smallest sided polygon in the n-uniform tiling.

We note that the appropriate angle of rotation in step (d) and scaling factor in step (e) are necessary for the locally flat foldable crease patterns to fold flat (see [15]).

Illustratio 1. As an example, consider the 3-uniform tiling $(3^6; 3^2.6^2; 6^3)$ shown in Fig. 1(a). There are three types of vertices in the tiling in which the tiles consisting of regular polygons meet. One vertex has six 3-gons (3^6), another vertex has two regular 3-gons and two regular 6-gons $(3^2.6^2)$, and the third vertex has three regular 6-gons (6^3). The dual tiling is shown in Fig. 1(b) with orange edges. It is obtained by forming edges connecting the incenters of every pair of adjacent tiles in $(3^6; 3^2.6^2; 6^3)$.

We first shrink the primal tiles, tiles of the 3-uniform tiling (Fig. 1(c)). Then we consider a dual tile of the dual tiling obtained by connecting the incenters of the primal tiles that meet in a vertex of the 3-uniform tiling. We insert the dual tile between the primal tiles by connecting each of its vertices to a vertex of a primal tile incident to a vertex of the 3-uniform tiling. Inserting every dual tile in the same way results to the pattern in Fig. 1(d). Performing steps (d) and

(e) give rise to the crease pattern shown in Fig. 1(e). In this example, we have $0 < \rho < \frac{\pi}{2} - \frac{\pi}{3} = \frac{\pi}{6}$.

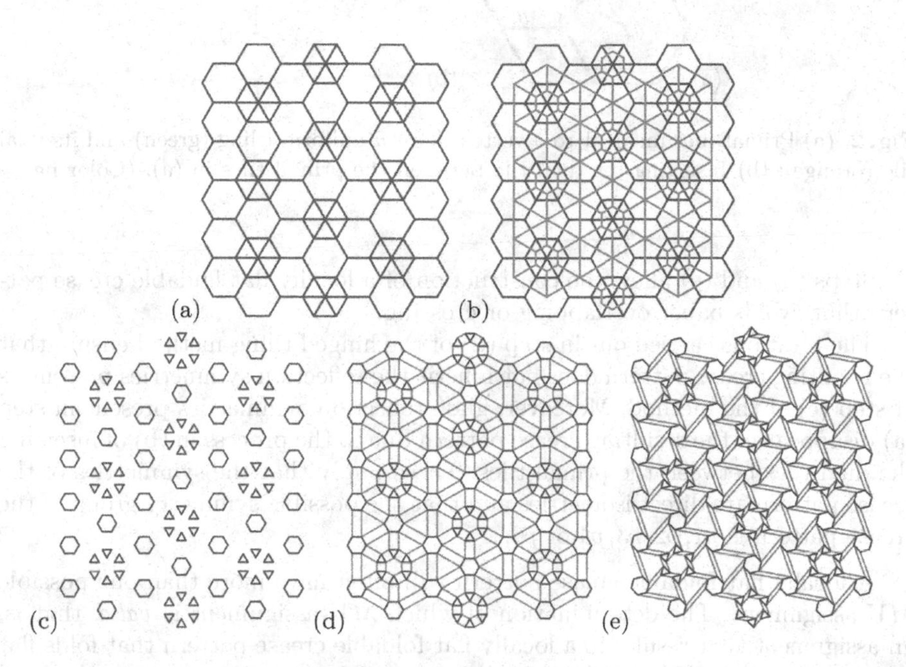

Fig. 1. Construction of a locally flat foldable crease pattern via the hinged tiling method.

Our result on locally flat foldable crease patterns is as follows.

Theorem 2. *There exists a locally flat foldable crease pattern from an n-uniform tiling using the hinged tiling method. The symmetry group of the crease pattern consists of direct isometries and is one of the following plane crystallographic groups: $p1, p2, p3, p4, p6$ (in IUCr notation [5]).*

Proof. In an n-uniform tiling, consider a vertex with the tiles (regular polygons) incident to it. Form a dual tile by connecting the incenters of these regular polygons. Note that the incircles of two adjacent regular polygons intersect at the midpoints of their common edges. Thus, the segments connecting the incenters of the adjacent polygons are perpendicular bisectors of their common edges. Two edges of the dual tile and two edges of a primal tile form a parallelogram $ABCD$ with $\angle B = \angle D = \frac{\pi}{2}$ (Fig. 2(a)). Thus $\angle A + \angle C = \pi$. Applying step (c) of the hinged tiling method results in a pattern where a vertex of a polygon has degree 4. In a vertex (Fig. 2(b)), we label the angles between the edges as $\alpha, \beta, \gamma, \delta$. Observe that $\alpha = \angle A$ and $\gamma = \angle C$. Thus $\alpha + \gamma = \pi$ and $\beta + \delta = \pi$, satisfying the Kawasaki's angle criterion.

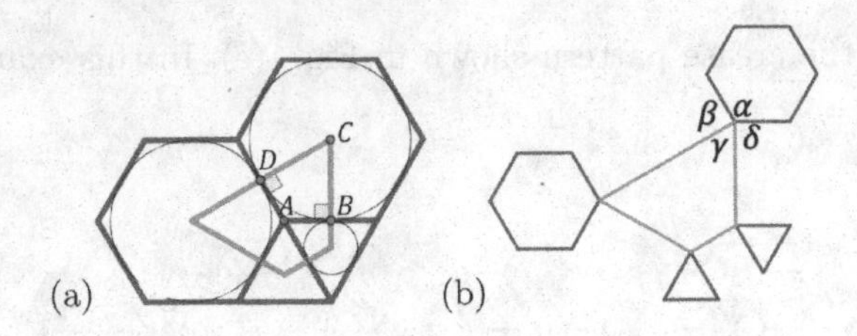

Fig. 2. (a) Primal tiles incident to a vertex of an n-uniform tiling (green) and its dual tile (orange); (b) Inserting the dual tile between the primal tiles in (a). (Color figure online)

Steps (d) and (e) allow the construction of a locally flat foldable crease pattern that avoids paper overlapping or crossing.

The rotations carried out in step (d) of the hinged tiling method ensure that the resulting crease pattern does not inherit the reflection symmetries present as of step (c) of the method. Moreover, glide reflection symmetries present in step (a) disappear in the resulting crease pattern due to the process in (b) of inserting the dual tiles between the primal tiles. These imply that the symmetries of the crease pattern are direct isometries and that the possible symmetry group of the crease pattern is $p1, p2, p3, p4$ or $p6$. □

A locally flat foldable crease pattern will often have more than one possible MV-assignment. The determination of which MV-assignment is *valid*, that is, an assignment that results to a locally flat foldable crease pattern that folds flat is guided by the following work of Maekawa and Justin.

Theorem 3. *For a flat-foldable single-vertex crease pattern with MV-assignment, the following hold:*

i. [8, 16] The number of M folds and the number of V folds differ by ± 2.
ii. [8, 9] If an angle θ_i is a strict local minimum (i.e., $\theta_{i-1} > \theta_i < \theta_{i+1}$ where θ_{i-1}, θ_i, θ_{i+1} are consecutive angles between the crease lines), then the two crease lines bounding angle θ_i have an opposite MV-assignment.

It follows from the above result that in a crease pattern obtained by the hinged tiling method, if a pair of adjacent edges of a parallelogram forming an angle greater than π is assigned with M folds, and the other pair of edges with V folds then a valid MV-assignment is obtained [10]. A particular assignment of M and V to a parallelogram that will give a valid MV-assignment to the crease pattern is called a *flat foldability type* of the parallelogram. There are two flat foldability types- one is obtained from the other through *crease inversion*, interchanging mountain and valley folds (Fig. 3). In the succeeding discussion we denote M and V folds with blue solid lines and red broken lines respectively.

3 Crystallographic Flat Origami

Consider a locally flat foldable crease pattern X obtained from an n-uniform tiling using the hinged tiling method. Suppose X has symmetry group that is a

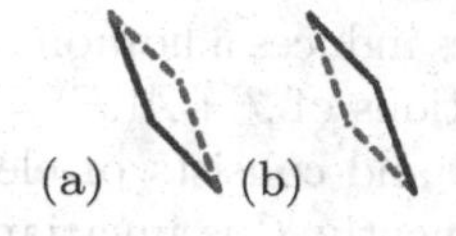

Fig. 3. Flat foldability types of a parallelogram.

plane crystallographic group G. We now discuss a systematic way of assigning flat foldability types to the set S of parallelograms in X, to obtain a locally flat foldable crease pattern $\mathcal{C}$ with MV-assignment which folds into a crystallographic flat origami. Observe that X is a union of sets of congruent parallelograms.

We proceed by considering a G-orbit $Gu = \{gu : g \in G\}$ of parallelograms, $u \in S$. Two parallelograms belong to one G-orbit if one can be sent to the other by an element in G. We take H, a subgroup of finite index in G, such that $Stab_H(v) = \{h \in H : hv = v\} = \{e\}$, v is any parallelogram in Gu. Consider $U = \{w_1, w_2, \ldots, w_n\}$ a set consisting of the H-orbit representatives in Gu. We have $Gu = HU = Hw_1 \cup Hw_2 \cup \cdots \cup Hw_n$.

Let $\mathcal{F} = \{f_1, f_2\}$ be the set of flat foldability types of a given parallelogram. We have f_1 and f_2 as shown in Fig. 3(a) and 3(b) respectively. The assignment of flat foldability types to the parallelograms in Gu will be treated as a partition $\mathcal{P}$ of Gu. Two parallelograms in $Gu = HU$ are assigned the same flat foldability types from $\mathcal{F}$ if and only if they belong to the same set in $\mathcal{P}$. Since we imposed the condition that $Stab_H(v) = \{e\}$ where v is a parallelogram in $Gu = HU$, then we are assured of no overlapping of MV-assignments in the parallelograms in Gu.

We consider one of the following partitions of Gu.

(i) $Gu = HU = \{hJU : h \in H\} = \{hJ\{w_1, w_2, \ldots, w_n\} : h \in H\}$ where $J < H$, $[H : J] = 2$, $h \in H \backslash J$.
(ii) $Gu = HU = Hw_1 \cup Hw_2 \cup \cdots \cup Hw_n = \{hJ_1w_1 : h \in H\} \cup \{hJ_2w_2 : h \in H\} \cup \cdots \cup \{hJ_nw_n : h \in H\}$ where $J_i < H$, $[H : J_i] = 2$, $i = 1, 2, \ldots, n$.

In (i), we partition the orbit $Gu = HU$ into a disjoint union of sets using one subgroup J of H. An alternative partition to Gu is when $G = H = J$, and one flat foldability type is assigned to Gu. In (ii) more than one subgroup of H is employed to partition the orbit $Gu = HU$. The process is repeated to the other G-orbits of parallelograms in S, until a locally flat foldable crease pattern $\mathcal{C}$ with MV-assignment is obtained. Arriving at several partitions of Gu (and thus several crease patterns with MV-assignments) entails determining the subgroup structures of G and H. The algorithm can best be implemented with the help of a computer software like GAP.

The method presented gives all possible crease patterns with MV-assignments whose symmetry group is a subgroup of finite index in G. The number of these crease patterns is finite, since there are a finite number of subgroups H in G of finite index.

Consider H^* a subgroup consisting of elements in G that either interchange or fix the flat foldability types in $\mathcal{F} = \{f_1, f_2\}$, $H \leq H^* \leq G$. The action of H^* on the set $\mathcal{F}$ of flat foldability types induces a homomorphism $\varphi : H^* \rightarrow A(\mathcal{F})$ where $A(\mathcal{F})$ is the group of permutations of $\mathcal{F} : A(\mathcal{F}) = \langle (f_1 \quad f_2) \rangle$. The kernel K^* of φ is a normal subgroup of H^* and consists of elements in H^* that fix the flat foldability types in $\mathcal{F}$. Consequently, $\mathcal{C}$ is invariant under K^* or has symmetry group K^*. The fundamental region of K^* in $\mathcal{C}$ corresponds to a fundamental region of a plane crystallographic group $\overline{K}$ in the flat origami such that $\overline{K} \cong K^*$. Thus, the flat origami is crystallographic. For the class of crystallographic flat origami discussed in this paper, we have $\overline{K} \cong p1, p2, p3, p4$, or $p6$.

The following results are helpful in determining the group consisting of all elements in G that fix the flat foldability types assigned to the parallelograms in a G-orbit. If there are m G-orbits then the symmetry group of $\mathcal{C}$ is given by $K^* = \cap_{i=1}^{m} K_i$, where K_i consists of all elements in G that fix the flat foldability types in a G-orbit Gu_i.

Theorem 4. *Suppose we have an assignment of flat foldability types to the parallelograms in a G-orbit Gu induced by the partition $Gu = Hw_1 \cup Hw_2 \cup \cdots \cup Hw_n = \{hJ_1w_1 : h \in H\} \cup \{hJ_2w_2 : h \in H\} \cup \cdots \cup \{hJ_nw_n : h \in H\}$, where $Stab_H(w_i) = \{e\}$, $J_i < H$, $[H : J_i] = 2$, $i = 1, 2, \ldots, n$. Then the group consisting of all elements that fix the flat foldability types is given by $K = J_1 \cap J_2 \cap \cdots \cap J_n$.*

Proof. From [18], we have $K = core_H(J_1 \cap J_2 \cap \cdots \cap J_n) = \cap_{h \in H} h(J_1 \cap J_2 \cap \cdots \cap J_n)h^{-1}$. Since every J_i is a normal subgroup of H, then $J_1 \cap J_2 \cap \cdots \cap J_n$ is a normal subgroup of H. Thus $K = J_1 \cap J_2 \cap \cdots \cap J_n$. □

In the next theorem, we assume that $\{e, y_2, y_3, \ldots, y_n\}$ is a complete set of right coset representatives of H in G. We have $H \leq H^* \leq G = H \cup Hy_2 \cup Hy_3 \cup \cdots \cup Hy_n$ so we can write $H^* = H \cup Hz_2 \cup Hz_3 \cup \cdots \cup Hz_r, r \leq n$, $z_2, z_3, \ldots, z_r \in \{e, y_2, y_3, \ldots, y_n\}$.

Theorem 5. *Suppose the assignment of flat foldability types to the parallelograms in $Gu = Hw_1 \cup Hw_2 \cup \cdots \cup Hw_n$ is induced by the partition $Gu = HU = \{hJ\{w_1, w_2, \ldots, w_n\} : h \in H\}$, where $Stab_H(w_i) = \{e\}$, $J < H$, $[H : J] = 2$. Then the group consisting of all elements that fix the flat foldability types is given by $K = \cap_{h \in H} h(J \cup Jz_2 \cup Jz_3 \cup \cdots \cup Jz_n)h^{-1}$. In particular,*

i. If $H^* = G$ then $K = \cap_{h \in H} h(J \cup Jy_2 \cup Jy_3 \cup \cdots \cup Jy_n)h^{-1}$.
ii. If $H^* = H$ then $K = J$.

If H^* is neither H nor G, one can determine K^* from among the normal subgroups of H^* using GAP [4].

Illustratio 2. We present three cases of crystallographic flat origami obtained from the locally flat foldable crease pattern X in Illustration 1, obtained from the 3-uniform tiling $(3^6; 3^2.6^2; 6^3)$. The symmetry group of X is $G = \langle a, x, y \rangle \cong p6$, where a is a $60°$ rotation centered at the center of a hexagon and x, y are linearly independent translations. There are three G-orbits of parallelograms in S: Gu_1 (violet), Gu_2(pink) and Gu_3(sky blue) (Fig. 4(a)). It will be shown that there are crystallographic flat origami that will arise from X, invariant under a group isomorphic to $p1, p2, p3, p6$.

A. We consider the simplest case of the partition where $G = H = J$ for each of the three G-orbits. That is, we assign one flat foldability type to each of Gu_1, Gu_2 or Gu_3. Note that $Stab_G(v) = \{e\}$, where v is a parallelogram in S.

If we assign the flat foldability type f_1(Fig. 3(a)) to each G-orbit of parallelograms, we obtain the locally flat foldable crease pattern $\mathcal{C}_1$ with MV-assignment as shown in Fig. 4(b). It has symmetry group G with fundamental region r_1 highlighted with color yellow.

If we assign the flat foldability type f_2 (Fig. 3(b)), we obtain the locally flat foldable crease pattern $\mathcal{C}_2$ with MV-assignment as shown in Fig. 4(c) with a fundamental region r_2 of its symmetry group G highlighted with yellow. Folding $\mathcal{C}_1$ produces the flat origami in Fig. 4(d), while folding $\mathcal{C}_2$ produces the flat origami in Fig. 4(e). The fundamental region r_1 of G corresponds to the fundamental region $\overline{r_1}$ of a group $\overline{G_1}$ in the flat origami. Similarly, the fundamental region r_2 corresponds to the fundamental region $\overline{r_2}$ of a group $\overline{G_2}$ in the flat origami. The resulting flat origami has symmetry group G, and is crystallographic. That is, $\overline{G_1} = \langle \overline{a}, \overline{x}, \overline{y} \rangle \cong G \cong p6$ and $\overline{G_2} = \langle \overline{a}, \overline{x}, \overline{y} \rangle \cong G \cong p6$ where $\overline{a}$ is a 60° rotation and $\overline{x}, \overline{y}$ are two linearly independent translations.

Other locally flat foldable crease patterns with MV-assignments will be obtained by assigning f_1 or f_2 to each of the three G-orbits Gu_1, Gu_2 or Gu_3. Table 1 summarizes the other six possibilities. A resulting crease pattern with MV-assignment has symmetry group isomorphic to G. It folds into a crystallographic flat origami with symmetry group $\overline{G} \cong G \cong p6$.

Table 1. Assignment of f_1 and f_2 to the G-orbits Gu_1, Gu_2, Gu_3 using partition (i) where $G = H = J$.

	Flat foldability type assignment			Crease pattern with	Folded pattern
	Gu_1	Gu_2	Gu_3	MV-assignment	
(1)	f_1	f_1	f_2	Fig. 5(a)	Fig. 5(g)
(2)	f_1	f_2	f_1	Fig. 5(b)	Fig. 5(h)
(3)	f_1	f_2	f_2	Fig. 5(c)	Fig. 5(i)
(4)	f_2	f_1	f_1	Fig. 5(d)	Fig. 5(j)
(5)	f_2	f_1	f_2	Fig. 5(e)	Fig. 5(k)
(6)	f_2	f_2	f_1	Fig. 5(f)	Fig. 5(l)

B. We still consider partition (i), where $G = H$, but this time we let $J_1 = J_2 = J_3 = \langle a^2, x, y \rangle \cong p3$ for each of the three G-orbits Gu_1, Gu_2 and Gu_3. The list of low index subgroups of $G = \langle a, x, y \rangle \cong p6$ given by GAP (see Table 2) suggests there is only one index 2 subgroup of G. Letting $J = \langle a^2, x, y \rangle$ we obtain the following partitions for each of the G-orbits: $Gu_1 = Ju_1 \cup aJu_1$, $Gu_2 = Ju_2 \cup aJu_2$ and $Gu_3 = Ju_3 \cup aJu_3$.

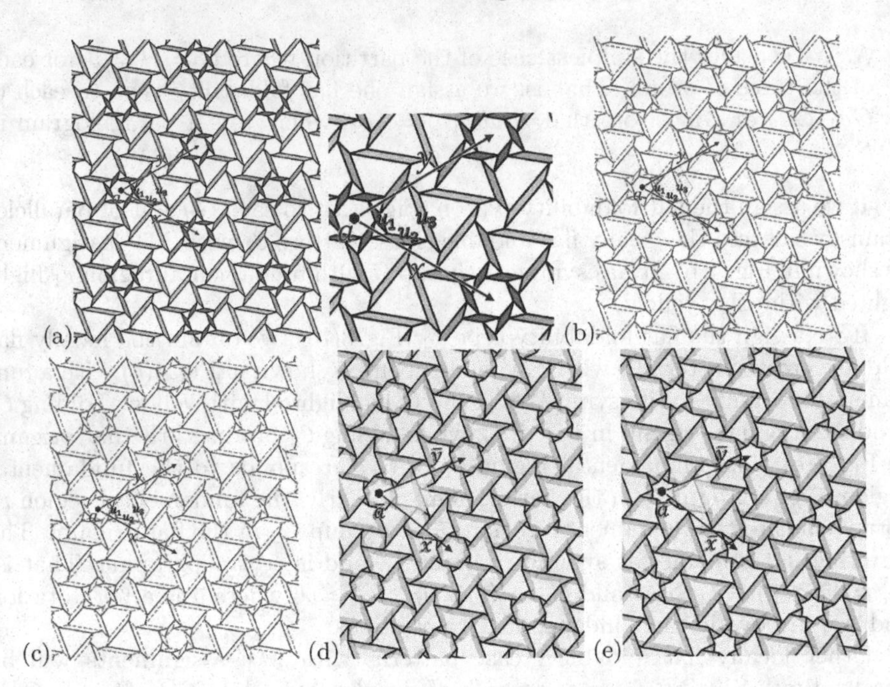

Fig. 4. (a) Crease pattern which is the orbit of $G = \langle a, x, y \rangle \cong p6$, showing the orbit representatives u_1, u_2, u_3. (b)-(c) MV-assignments where each parallelogram is assigned with only one flat foldability type of the parallelogram. (d)-(e) Corresponding folded patterns of the MV-assignments in (b) and (c).

Table 2. Subgroups of $G = \langle a, x, y \rangle \cong p6$ of small index generated using GAP.

Subgroup H	$[G : H]$	Subgroup H	$[G : H]$	Subgroup H	$[G : H]$
$\langle a, x, y \rangle \cong p6$	1	$\langle a, x^2, y^2 \rangle \cong p6$	4	$\langle ya, x, y^2 \rangle \cong p6$	6
$\langle a^2, x, y \rangle \cong p3$	2	$\langle x, y \rangle \cong p1$	6	$\langle a^2, xy, yx \rangle \cong p3$	6
$\langle a, xy, yx \rangle \cong p6$	3	$\langle a^3, x, y^2 \rangle \cong p2$	6	$\langle xa^{-2}, xy, yx \rangle \cong p3$	6
$\langle a^3, x, y \rangle \cong p2$	3				

The assignment of flat foldability types giving rise to different locally flat foldable crease patterns with MV-assignment are shown in Table 3. One possibility is to assign the flat foldability type f_1 to Ju_i and f_2 to aJu_i, $i = 1, 2, 3$ (Table 3 (1)). This assignment results in the locally flat foldable crease pattern with MV-assignment presented in Fig. 6(a). For every G-orbit in the crease pattern with MV-assignment, the elements in G interchange or fix the flat foldability types, and J is the largest subgroup of G that consists of elements that

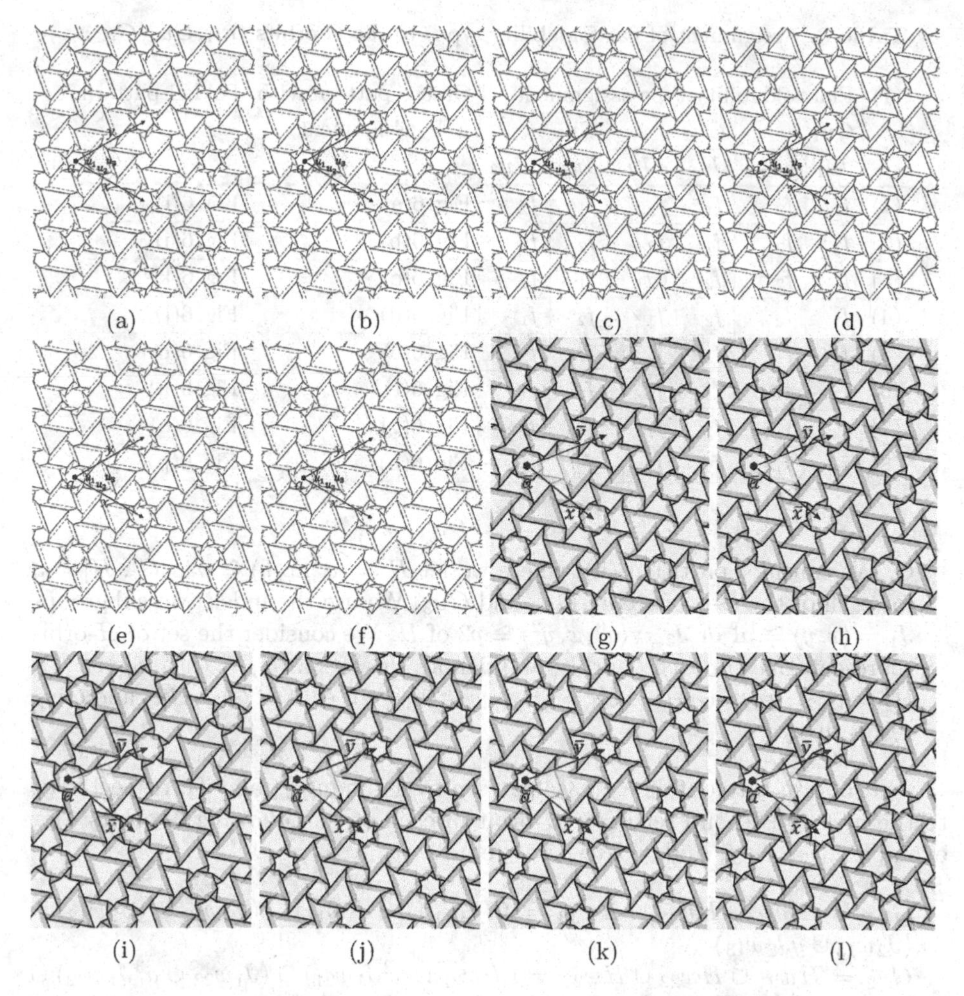

Fig. 5. Crease patterns with MV-assignments and their corresponding folded patterns.

fix the flat foldability types. Thus, the crease pattern with MV-assignment has symmetry group J. A fundamental region of J in the crease pattern with MV-assignment corresponds to a fundamental region of $\overline{J} \cong J \cong p3$ in the folded pattern. The crystallographic flat origami, shown in Fig. 6(i), has symmetry group $\overline{J} \cong J \cong p3$. The other crease patterns with MV-assignment (Table 3 (2)–(8)) also has symmetry group $J \cong p3$. These crease patterns fold to crystallographic flat origami with symmetry group $\overline{J} \cong J \cong p3$.

Table 3. Assignment of flat foldability types to the G-orbits Gu_1, Gu_2, Gu_3.

(1)	Flat foldability type assignment						Crease pattern with MV-assignment	Folded pattern
	Gu_1		Gu_2		Gu_3			
	Ju_1	aJu_1	Ju_2	aJu_2	Ju_3	aJu_3		
(1)	f_1	f_2	f_1	f_2	f_1	f_2	Fig. 6(a)	Fig. 6(i)
(2)	f_1	f_2	f_1	f_2	f_2	f_1	Fig. 6(b)	Fig. 6(j)
(3)	f_1	f_2	f_2	$6f_1$	f_1	f_2	Fig. 6(c)	Fig. 6(k)
(4)	f_1	f_2	f_2	f_1	f_2	f_1	Fig. 6(d)	Fig. 6(l)
(5)	f_2	f_1	f_1	f_2	f_1	f_2	Fig. 6(e)	Fig. 6(m)
(6)	f_2	f_1	f_1	f_2	f_2	f_1	Fig. 6(f)	Fig. 6(n)
(7)	f_2	f_1	f_2	f_1	f_1	f_2	Fig. 6(g)	Fig. 6(o)
(8)	f_2	f_1	f_2	f_1	f_2	f_1	Fig. 6(h)	Fig. 6(p)

C. We consider partition (ii), where we let $H = \langle a^3, x, y\rangle \cong p2$, $[G : H] = 3$ for each of the G-orbits Gu_1, Gu_2 and Gu_3. We use the index two subgroups $J_1 = \langle x, y\rangle \cong p1$ or $J_2 = \langle a^3, x, y^2\rangle \cong p2$ of H. We consider the set of H-orbit representatives in each G-orbit, and denote the sets as $U_1 = \{w_{11}, w_{12}, w_{13}\}$, $U_2 = \{w_{21}, w_{22}, w_{23}\}$ and $U_3 = \{w_{31}, w_{32}, w_{33}\}$ for each of Gu_1, Gu_2 and Gu_3, respectively.

One option is to obtain a crease pattern by assigning f_1 and f_2, respectively, to the first and second set in an H-orbit in each of the following G-orbits: (for example J_1w_{11} is f_1 and $a^3J_1w_{11}$ is f_2 etc.)

$Gu_1 = Hw_{11} \cup Hw_{12} \cup Hw_{13} = (J_1w_{11} \cup a^3J_1w_{11}) \cup (J_2w_{12} \cup yJ_2w_{12}) \cup (J_2w_{13} \cup yJ_2w_{13})$
$Gu_2 = Hw_{21} \cup Hw_{22} \cup Hw_{23} = (J_1w_{21} \cup a^3J_1w_{21}) \cup (J_1w_{22} \cup a^3J_1w_{22}) \cup (J_2w_{23} \cup yJ_2w_{23})$
$Gu_3 = Hw_{31} \cup Hw_{32} \cup Hw_{33} = (J_1w_{31} \cup a^3J_1w_{31}) \cup (J_1w_{32} \cup a^3J_1w_{32}) \cup (J_1w_{33} \cup a^3J_1w_{33})$.

This results in the crease pattern with MV-assignment shown in Fig. 7(a). The elements in $H = \langle a^3, x, y\rangle \cong p2$ either fix or interchange the flat foldability types in each G-orbit.

The group K_1 and K_2 consisting of elements in H that fix the flat foldability types in Gu_1 and Gu_2 respectively is $K_1 = K_2 = J_1 \cap J_2 = \langle x, y^2\rangle \cong p1$. The group K_3 consisting of elements in H that fix the flat foldability types is Gu_3 is $K_3 = J_1 = \langle x, y\rangle \cong p1$. In the crease pattern with MV-assignment, the elements in $K = K_1 \cap K_2 \cap K_3 = \langle x, y^2\rangle \cong p1$ fix the flat foldability types. The resulting crystallographic flat origami (Fig. 7(b)) is invariant under the plane crystallographic group $\overline{K} \cong K \cong p1$.

Another crease pattern with MV-assignment can be obtained by choosing different combinations of index 2 subgroups of $H = \langle a^3, x, y\rangle \cong p3$, to be used

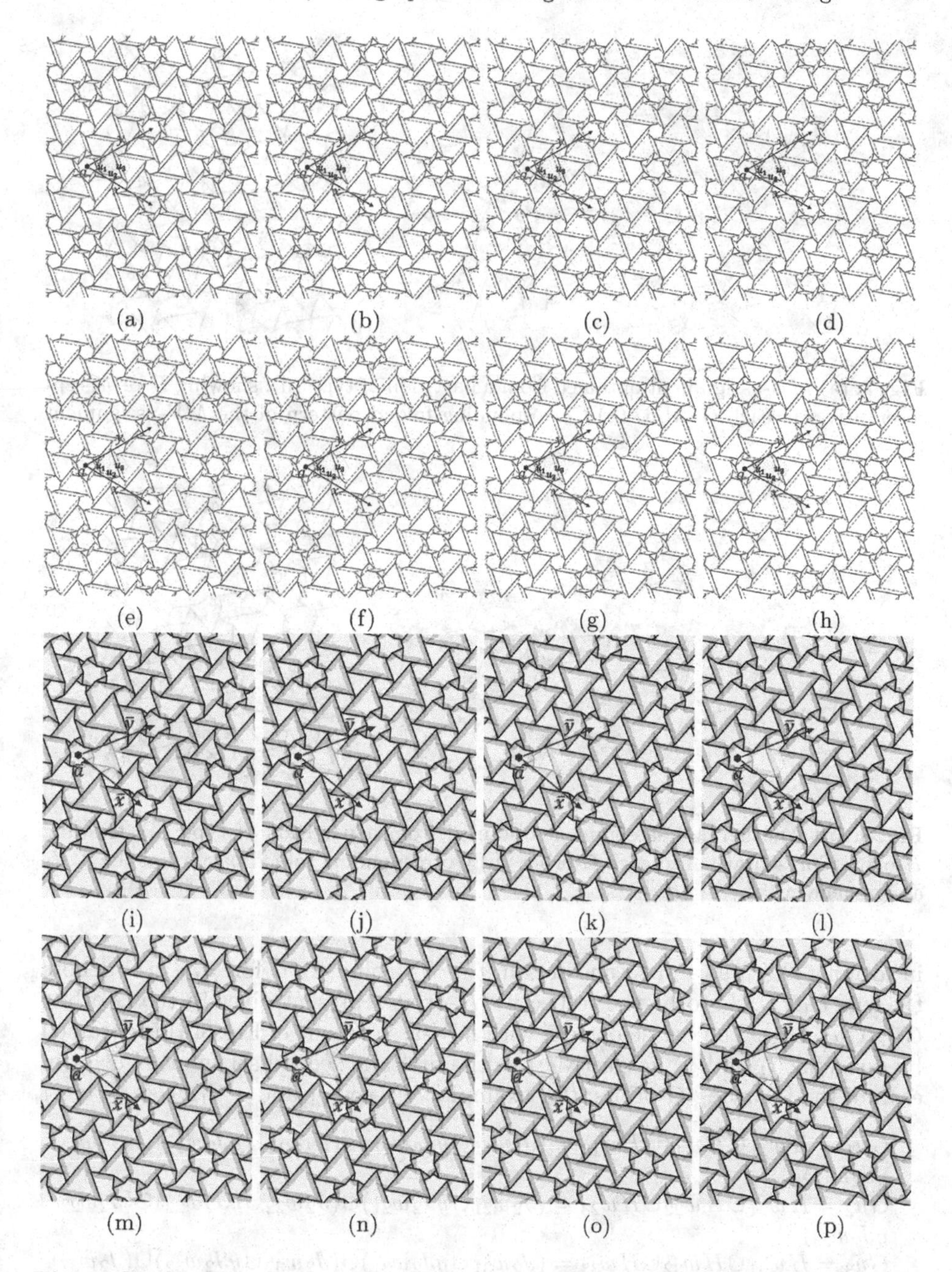

Fig. 6. Crease patterns with MV-assignments corresponding to the partition $X = JU \cup aJU$ where $J = \langle a^2, x, y\rangle \cong p3$, $[H : J] = 2$.

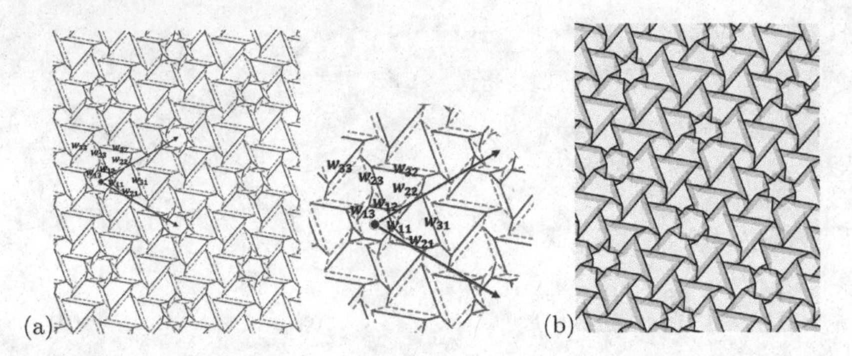

Fig. 7. (a) MV-assignment obtained from a partition of type (ii), showing H-orbit representatives in each G-orbit. (b) Corresponding folded pattern of the MV-assignment in (a).

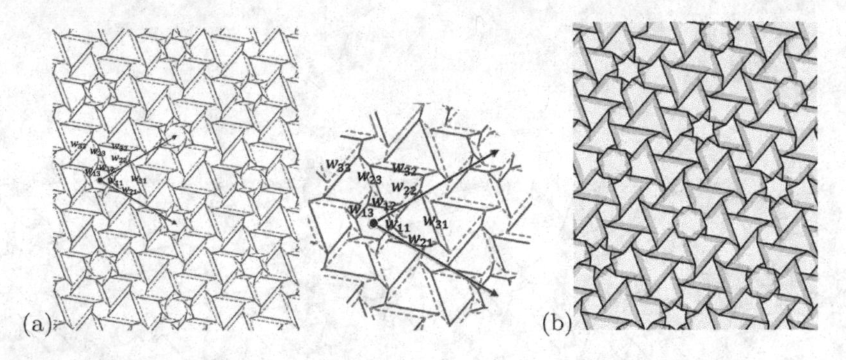

Fig. 8. (a) Another MV-assignment obtained from a partition of type (ii), showing H-orbit representatives in each G-orbit. (b) Corresponding folded pattern of the MV-assignment in (a).

in forming the first and second sets in the H-orbits. As before, we still denote the sets U_1, U_2, U_3 to be the respective H-orbit representatives in the G-orbits Gu_1, Gu_2, Gu_3. The crease pattern with MV-assignment in Fig. 8(a) is obtained by assigning f_1 and f_2, respectively, to the first and second set in an H-orbit in each of the following G-orbits: (for example $J_1 w_{11}$ is f_1 and $a^3 J_1 w_{11}$ is f_2 etc.)

$Gu_1 = Hw_{11} \cup Hw_{12} \cup Hw_{13} = (J_2 w_{11} \cup y J_2 w_{11}) \cup (J_2 w_{12} \cup y J_2 w_{12}) \cup (J_2 w_{13} \cup y J_2 w_{13})$
$Gu_2 = Hw_{21} \cup Hw_{22} \cup Hw_{23} = (J_2 w_{21} \cup y J_2 w_{21}) \cup (J_2 w_{22} \cup y J_2 w_{22}) \cup (J_2 w_{23} \cup y J_2 w_{23})$
$Gu_3 = Hw_{31} \cup Hw_{32} \cup Hw_{33} = (J_2 w_{31} \cup y J_2 w_{31}) \cup (J_2 w_{32} \cup y J_2 w_{32}) \cup (J_2 w_{33} \cup y J_1 w_{33})$.

In this crease pattern with MV-assignment we have $H^* = H = \langle a^3, x, y \rangle \cong p2$. In each G-orbit, the elements in H that fix the flat foldability types comprise $J_2 = \langle a^3, x, y^2 \rangle \cong p2$, thus $K^* = J_2 \cong p2$. Folding this crease pattern produces the crystallographic flat origami (Fig. 8(b)) with symmetry group $\overline{K} \cong K^* \cong p2$.

4 Iso-Area Crystallographic Flat Origami

The term *iso-area* was coined by Kawasaki (see [13]) to mean an origami where equal portions of each side of the paper are shown on the exterior of the origami. Maekawa [16] redefined the term to refer to origami with crease patterns exhibiting symmetry that is unchanged under an n-fold rotation plus a crease inversion (interchanging mountain and valley folds). Lang [14] redefined the term further to refer to an origami whose crease pattern with MV-assignment remains the same, after interchanging mountain and valley folds and performing some combination of translation, rotation, reflection, inversion, and/or magnification. The crystallographic flat origamis in Fig. 6 are examples of iso-area origami. We have the following result on obtaining iso-area crystallographic flat origami from the method discussed in the previous section.

Theorem 6. *Consider a crease pattern X with symmetry group G obtained from a n-uniform tiling via the hinged tiling method. If a locally foldable crease pattern with valid MV-assignment is represented by the partition $JU \cup hJU$, where $h \in H \backslash J$, $J < H \leq G$, $[H : J] = 2$, and U is a complete set of H-orbit representatives of the sets of parallelograms in X, then the resulting crystallographic flat origami is an iso-area.*

Proof. Consider a crease pattern with MV-assignment $\mathcal{C}$ represented by the partition $JU \cup hJU$, $h \in H \backslash J$, $[H : J] = 2$. Let $\mathcal{F} = \{f_1, f_2\}$ be the set of flat foldability types assigned to elements in S. Suppose JU is assigned with f_1 and hJU, $h \in H \backslash J$ is assigned with f_2. Recall that f_1 and f_2 are crease inversions of each other, that is one can be obtained from the other by interchanging mountain and valley folds. Assigning f_2 to JU and f_1 to hJU results in a crease pattern $\mathcal{C}^*$ with MV-assignment that is a crease inversion of $\mathcal{C}$. Note that applying h^{-1} to the partition $hJU \cup JU$ will yield $h^{-1}(hJU \cup JU) = JU \cup h^{-1}JU = JU \cup hJU$. Thus, the crystallographic flat origami is an iso-area. □

5 Conclusion and Outlook

In this paper, we apply the hinged tiling method to an n-uniform tiling to arrive at a locally flat foldable crease pattern X. This crease pattern has symmetry group G consisting only of direct isometries. We presented a method on how to systematically assign valid MV-assignments to X to arrive at crystallographic flat origami. The method presented would give all possible crease patterns with MV-assignments with symmetry group a subgroup of finite index in G. Depending on the assignment of flat foldability types to parallelograms of X using either the partition (i) or (ii), we obtain a locally flat foldable crease pattern $\mathcal{C}$ with MV-assignment having symmetry group K^*. The group K^* is a normal subgroup of H^*, and is the group consisting of all the elements in G that interchange or fix the flat foldability types in $\mathcal{C}$. In particular, the elements in K^* fix the flat foldability types and form the symmetry group of $\mathcal{C}$. The crystallographic flat

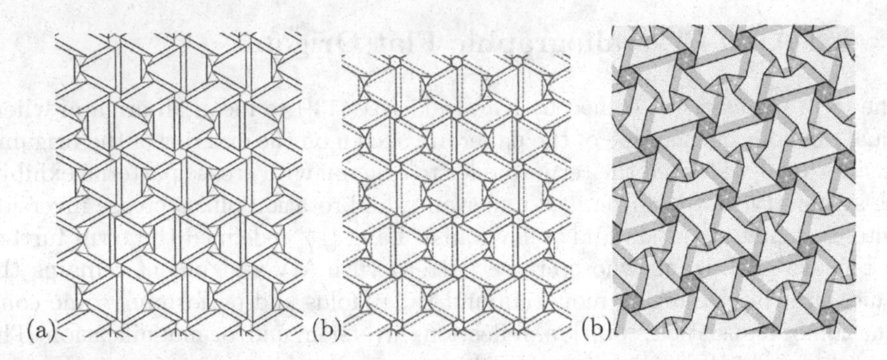

Fig. 9. (a) Offset twist crease pattern from tiling $(3^26^2; 3.6.3.6)$. (b) An *MV*-assignment of crease pattern in (a) that folds to (c) crystallographic flat origami.

origami obtained from $\mathcal{C}$ is invariant under $\overline{K} \cong K^*$. Herein we gave some examples of crystallographic flat origami arising from a particular locally flat foldable crease pattern using the method discussed. The method we present here could give rise to algorithms which could be implemented via a computer program, and will help facilitate an easier listing of all the possible crystallographic flat origami from one locally flat foldable crease pattern. We propose this as a next step in the study.

We also suggest to investigate symmetry groups of crystallographic flat origami obtained from offset twist crease patterns [14] arising from n-uniform tilings. An example of an offset twist crease pattern obtained from the 2-uniform tiling $(3^26^2; 3.6.3.6)$ is given in Fig. 9(a). Its symmetry group is $G' = \langle a, b, x, y \rangle \cong p2mg$ where a is a 180° rotation centered at a parallelogram, b is a reflection with horizontal axis, and x, y are two linearly independent translations. Its symmetry group contains isometries that are not direct isometries. It can be considered as union of orbits of trapezoids under G. One can show that a trapezoid in this crease pattern has also two flat foldability types. A flat foldability type of the trapezoid is such that the longest base is assigned with mountain fold and the other edges with valley folds. The other flat foldability is the crease inversion of the first one.

The crease pattern in Fig. 9(b) is obtained by assigning with one flat foldability type each orbit of trapezoids under $G' \cong p2mg$. Folding these crease pattern results to the crystallographic flat origami with symmetry group $\overline{J} \cong p2mg$.

It would also be interesting to determine conditions for offset twist crease patterns to produce iso-area crystallographic flat origami.

References

1. Bateman, A.: Computer tools and algorithms for origami tessellation design. In: Hull, T. (ed.) Origami^3: Third International Meeting of Origami Science, Math, and Education, pp. 121–127. CRC Press, Florida (2002)

2. Bern, M., Hayes, B.: The complexity of flat origami. In: Proceedings of the Seventh Annual ACM-SIAM Symposium on Discrete Algorithms, pp. 175–183. Atlanta, Georgia (1996)
3. De Las Peñas, M. L. A., Taganap, E., Rapanut, T.: Color symmetry approach to the construction of crystallographic flat origami. In: Miura, K., et al. (eds.) Origami 6: I. Mathematics, pp. 11–20 (2015)
4. The GAP Group, GAP - Groups, Algorithms, and Programming, Version 4.7.7 (2015). (http://www.gap-system.org)
5. Hahn, T. (ed.): International tables for crystallography, Vol. A, Space-Group Symmetry, 5th ed. Springer, Heidelberg (2005)
6. Hull, T.: On the mathematics of flat origamis. Congr. Numer. **100**, 215–224 (1994)
7. Hull, T.: Counting Mountain-valley assignments for flat folds. Ars Comb. **67**, 175–188 (2003)
8. Justin, J.: Toward a mathematical theory of origami. In: Miura, K. (ed.) Origami Science and Art: Proceedings of the Second International Meeting of Origami Science and Scientific Origami, pp. 15–29, Seian University of Art and Design, Otsu, Japan (1997)
9. Kawasaki, T.: On the relation between mountain-creases and valley-creases of a flat origami (abridged English translation) In: Huzita, H. (ed.) Proceedings of the First International Meeting of Origami Science and Technology, pp. 229–237. Ferrara, Italy (1989)
10. Kawasaki, T.: Roses, Origami & Math. Japan Publications, Tokyo, Japan (2005)
11. Kawasaki, T., Yoshida, M.: Crystallographic flat origamis. Mem. Fac. Sci. Kyushu Univ. Ser. A **42**(2), 153–157 (1988)
12. Kasahara, K.: Origami Omnibus, p. 96. Japan Publications, Tokyo, Japan (1988)
13. Kasahara, K., Takahama, T.: Origami for the Connoisseur, p. 29. Japan Publications, Tokyo (1987)
14. Lang, R.: Twists, Tilings and Tessellations: Mathematical Methods for Geometric Origami, pp. 236, 287. CRC Press, Florida, USA (2018)
15. Lang, R., Bateman, A.: Every spider web has a simple flat twist tessellation. In: Wang-Iverson, P., et al. (eds.) Origami[5] : Fifth International Meeting of Origami Science. Mathematics and Education, pp. 455–474. CRC Press, Florida (2010)
16. Maekawa, J.: The definition of Iso-Area folding. In: Hull, T. (ed.) Origami[3]: Proceedings of the Third International Meeting on Origami Science, Mathematics and Education, pp. 53–58. A.K. Peters Ltd., Natick, MA (2002)
17. Sales, R.E.R.: On the Crystallographic Flat Origami. Ph.D. Dissertation, University of the Philippines Diliman, October 2000
18. Say-awen, A. L., De Las Peñas, M. L. A., Rapanut, T.: On color fixing groups associated with colored symmetrical tilings. In: Proceedings of the 3rd International Conference on Mathematical Sciences, pp. 050012-1-050012–10 (2014)
19. Taganap, E., De Las Peñas, M. L. A., Rapanut, T.: Crystallographic flat origami with three flat foldability types of the generating unit. In: Proceedings of the 3rd International Conference on Mathematical Sciences, pp. 662–667 (2014)
20. Verrill, H.: Origami tessellations. In: Sarhangi, R. (ed.) Conference Proceedings of Bridges: Mathematical Connections in Art, Music and Science, pp. 55–68 (1998)
21. Wang-Iverson, P., Lang, R., Yang, M. (eds.): Origami[5]: Fifth International Meeting of Origami Science. Mathematics and Education, pp. 455–474. CRC Press, Florida (2010)

Packing Cube Nets into Rectangles with $O(1)$ Holes

Erik D. Demaine[1], Martin L. Demaine[1], Ryuhei Uehara[2], Yushi Uno[3(✉)], and Andrew Winslow[4]

[1] Massachusetts Institute of Technology, Cambridge, MA, USA
{edemaine,mdemaine}@mit.edu
[2] Japan Advanced Institute of Science and Technology, Nomi, Ishikawa, Japan
uehara@jaist.ac.jp
[3] Osaka Prefecture University, Sakai, Osaka, Japan
uno@cs.osakafu-u.ac.jp
[4] University of Texas Rio Grande Valley, Edinburg, TX, USA

Abstract. We show that the 11 hexomino nets of the unit cube (using arbitrarily many copies of each) can pack disjointly into an $m \times n$ rectangle and cover all but a constant c number of unit squares, where $4 \leq c \leq 14$ for *all* integers $m, n \geq 2$. On the other hand, the nets of the dicube (two unit cubes) can be exactly packed into some rectangles.

Keywords: Unfoldings · Packings · Polyominoes · Hexominoes

1 Introduction

Packing with Polyominoes. Polyominoes[1] have attracted the attention of mathematicians, computer scientists, and amateur researchers since their invention by Solomon Golomb in 1954 [9,11] and popularization by Martin Gardner since 1959 [7,8]. Since this early beginning, the primary focus has been to study packing of polyominoes into "nice" shapes. For example, Golomb's original paper [9] studies which subsets of the 8×8 checkerboard can and cannot be exactly packed with dominoes, trominoes, and/or tetrominoes, possibly with a small number of monominoes. Gardner's original article [7] includes a proof by Golomb that the 35 hexominoes (each used exactly once) cannot exactly pack any rectangle, even though there are six rectangles with the proper area. Gardner wrote: "I seriously considered offering $1,000 to the first reader who succeeded in constructing one of these six rectangles, but the appalling thought of hours that might be wasted on the challenge forced me to relent". In this spirit, the Eternity puzzle is a 209-piece polyomino-like packing puzzle that awarded £1 million to the first solvers, Alex Selby and Oliver Riordan, one year after its release [16,20].

[1] An *n-omino* or *polyomino* is an edge-to-edge joining of n unit squares. The special cases $n = 1, 2, 3, 4, 5, 6$ are called *monominoes*, *dominoes*, *trominoes*, *tetrominoes*, *pentominoes*, and *hexominoes*, respectively.

J. Akiyama et al. (Eds.): JCDCGGG 2018, LNCS 13034, pp. 152–164, 2021.
https://doi.org/10.1007/978-3-030-90048-9_12

Packing with Hexominoes. In this paper, we study packing of *hexominoes* into *rectangles*. In 1966, Golomb [10] proved that nine hexomino shapes can each exactly pack some finite rectangle (by using arbitrarily many copies), while 25 hexomino shapes can each exactly pack infinite forms of rectangles (three-sided half-strips, four-sided Ls, two-sided strips, or the entire plane) but cannot exactly pack any finite rectangle. As described by Gardner [8], these results left unsolved exactly one hexomino shape (of 35); twenty-one years later, blind software engineer Dahlke [4] used computer search to find an exact packing of this hexomino shape into the smallest possible rectangle, 23×24. Every exact packing of hexominoes into some finite rectangle leads to exact packings into infinitely many finite rectangles (by repetition). Klarner [14] studied which rectangles can be exactly packed by a single polyomino shape. Bos [2] gave several exact packings of squares by two hexomino shapes (using arbitrarily many copies of each). See the extensive bibliography on "rectifiable polyominoes" [18]. In the modern Internet era, many websites are devoted to exact packings of nice shapes by various sets of hexominoes [3,5,6,17,18,21].

Cube Nets. We focus on the eleven hexominoes that fold into a unit cube, as shown in Fig. 1. More precisely, an *edge unfolding* of a cube is a way to cut some edges and unfold the remaining edges to produce a flat nonoverlapping polygon, called a *net*. Only eleven of the 35 hexominoes are nets of the cube. For convenience, we name them alphabetically from A to K (different from the usual labeling of hexominoes).

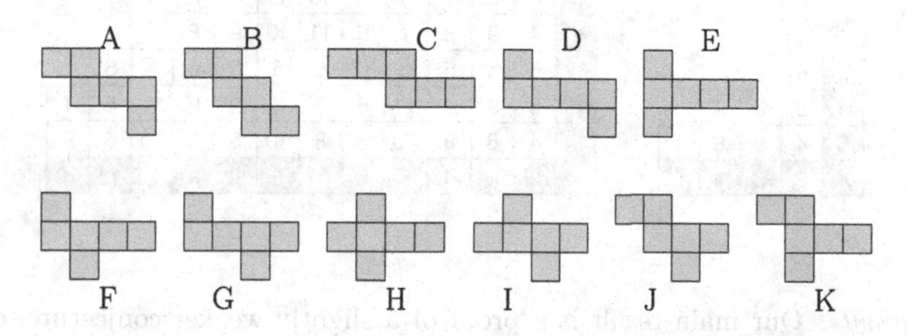

Fig. 1. Eleven cube nets labeled from A to K (as in [12]).

What about packings by cube nets? A recent paper [1] considered exact packing of cube nets (using arbitrarily many copies of each) onto the surface of a (larger) cube. For packing into rectangles, none of the eleven cube nets are among the ten hexomino shapes that by themselves can exactly pack a rectangle. More surprising is that no combination of cube nets can exactly pack a rectangle:

Theorem 1 (Folklore, proved in [12,13]**).** *No rectangle can be exactly packed by nets of the cube.*

Near-Exact Packing. Therefore, we relax the goal, and aim instead for "near-exact" packings of cube nets into various rectangles. In other words, we aim to find packings that have no overlap and cover all but a few 1×1 *holes* or *uncovered cells*. Specifically, we give nearly tight bounds on the following problem:

> For all positive integers m, n, what is the maximum number of cube nets that can be packed into an $m \times n$ rectangle. Equivalently, what is the least number of uncovered cells that can remain?

This specific problem has been studied extensively for small m, n. It is known, as folklore, that the minimum-area rectangle into which all 11 nets can pack is $11 \times 7 = 77$. In perhaps the first work on the problem, Odawara [15] presented several packings and gave a table of results that he could achieve up to 12×12. Intriguingly, he conjectured that every rectangle with $m, n \geq 7$ can have all but between 6 and 11 squares covered by packing cube nets. Inoha et al. [12] improved and extended these results by exhaustive computer search using BurrTools [19]. Table 1 summarizes all of the results now known to be optimal.

Table 1. Known optimal values of uncovered cells in each size $m \times n$ of rectangles for $2 \leq n \leq m \leq 14$. The value is 12 only for rectangles 6×6 and 12×6.

2	4			
3	6	9		
4	8	6	4	
5	4	9	8	7
×	2	3	4	5

6	6	6	6	6	12							
7	8	9	4	11	6	7						
8	4	6	8	10	6	8	10					
9	6	9	6	9	6	9	6	9				
10	8	6	4	8	6	10	8	6	10			
11	4	9	8	7	6	11	10	9	8	7		
12	6	6	6	6	12	6	6	6	6	6	6	
13	8	9	4	11	6	7	8	9	10	11	6	7
14	4	6	8	10	6	8	10	6	8	10	6	8
×	2	3	4	5	6	7	8	9	10	11	12	13

Our Results. Our main result is a proof of a slightly weaker conjecture: every $m \times n$ rectangle with $m, n \geq 2$ can be packed by cube nets leaving at most 14 uncovered cells (Sect. 2). ($1 \times n$ rectangles cannot fit any cube net from Fig. 1, so these are all the rectangles of interest.) Such a worst-case upper bound cannot be improved beyond 12, as the 6×6 and 6×12 rectangles require that many uncovered cells, so our upper bound of 14 is close to tight. We also prove a stronger form of Theorem 1: every $m \times n$ rectangle with $m, n \geq 2$ must leave at least 4 uncovered cells (Sect. 3). This best-case lower bound is tight because 2×8, 4×7, etc. rectangles can be packed with just 4 uncovered cells. In contrast, we show that the nets of a dicube (two cubes glued face-to-face) behave very differently: they admit an exact packing of some rectangle, and thus infinitely many rectangles (Sect. 4).

BurrTools. BurrTools [19] is a powerful software tool for exhaustively exploring packings of a specified set of shapes into a specified shape (e.g., a rectangle). Figure 2 shows a snapshot of the software. We used BurrTools extensively on constant-size instances to search for patterns in the packings which we then generalized by hand into the infinite family of packings presented in Sect. 2.

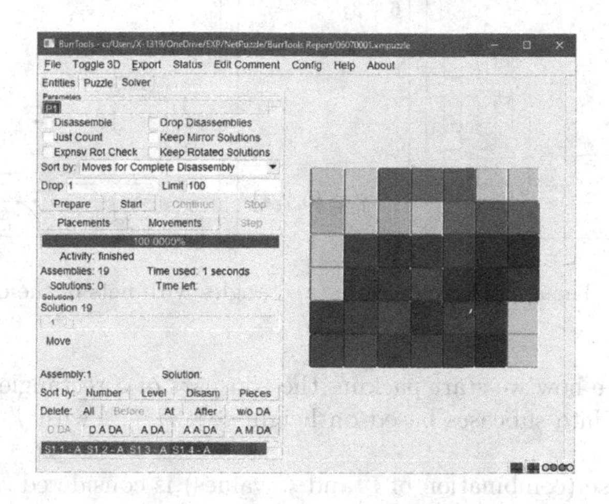

Fig. 2. A snapshot of BurrTools.

2 Upper Bound on Uncovered Cells

In this section, we prove the following theorem, providing upper bounds for the worst-case number of uncovered cells when we pack nets of the cube in a rectangle.

Theorem 2. *Let $m, n \in \mathbb{N}$. If $m \geq n \geq 2$, then nets of the cube can pack an $m \times n$ rectangle leaving at most 14 uncovered cells.*

We prove this theorem by constructing packing patterns that satisfy the stated condition. We split into four cases according to the size $m \times n$ of the rectangle: (i) $\{\geq 8\} \times \{\geq 6\}$ (i.e., 8×6 or larger); (ii) $\{6, 7\} \times \{6, 7\}$; (iii) $\{\geq 6\} \times \{2, 3, 4, 5\}$; (iv) $\{2, 3, 4, 5\} \times \{2, 3, 4, 5\}$ (i.e., 5×5 or smaller). Each of these cases corresponds to one of the subsequent lemmas.

Lemma 1. *Rectangles of dimensions $m \times n$ with $m \geq 8$ and $n \geq 6$ (and $m \geq n$) can be packed with nets of the cube leaving at most 14 cells uncovered.*

Proof. The proof is constructive, broken into several cases based on small modifications to two general packing formats seen in Fig. 3. The two top-level cases are rectangles of even ($6i + i'$ with $i' \in \{0, 2, 4\}$) and odd ($6i + i'$ with $i' \in \{1, 3, 5\}$) width. Note that this combination can realize any width of 6 or larger. These

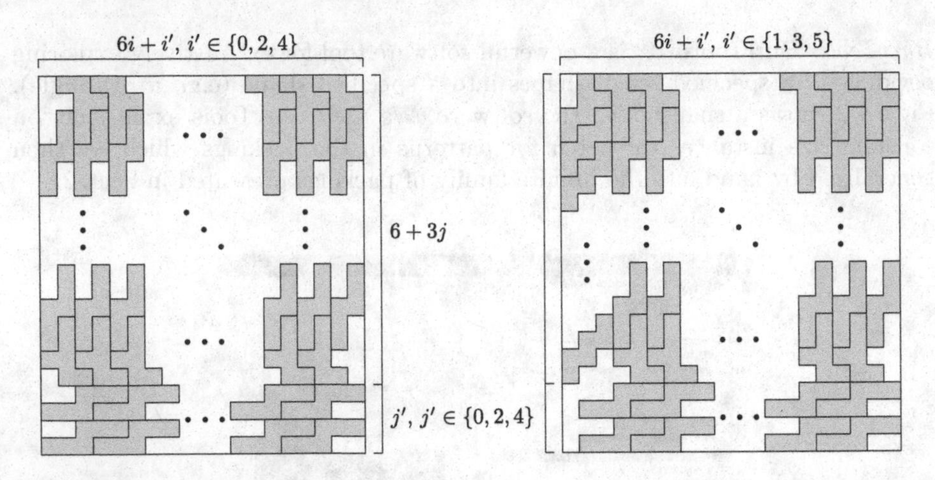

Fig. 3. The approach to packing rectangles with nets of the cube.

figures illustrate how we start packing the top part of a rectangle. Each case is further broken into subcases based on height $6 + 3j + j'$ with j' equal to 0, 2, or 4.

Each subcase (combination of i' and j' values) is considered separately. For compactness, the arrangement of the upper portion of each packing is excluded unless it does not agree with the arrangement seen in Fig. 3. Figures 4 and 5 contain the subcases with $i' \in \{0, 2, 4\}$ and $i' \in \{1, 3, 5\}$, respectively.

The vertical dimension is assumed to be equal to $6 + 3j + j'$ with $j \in \{0, 2, 4, \dots\}$ and $j' \in \{0, 2, 4\}$. Note again that any integer at least 8 can be written in such a form. Inspection of the arrangement of each subcase is sufficient to observe that every subcase accommodates all values of $i \geq 1$ (i.e., widths of $6i + i'$ for some subset of i' and all $i \geq 1$).

In the even-width case, we first see that there are 6 cells which remain uncovered in the top region (Fig. 3, left), so it suffices to show that we can cover the other (especially the bottom) region with at most 8 cells remaining uncovered. We can confirm this fact in Fig. 4, except the case when the width is $6i + i'$ with $i' = 4$ and the height is $6 + 3j + j'$ with $j' = 4$. This case is considered separately, and is covered with 10 uncovered cells in total (Fig. 4, bottom right).

In the odd-width case, we see similarly that there are 3 cells remaining uncovered in the top region (Fig. 3, right), which implies that it suffices to show that we can cover the other region with at most 11 cells remaining uncovered. We can confirm all the cases in Fig. 5. This completes the proof. □

Lemma 2. *Rectangles of dimensions $m \times n$ with $7 \geq m \geq n \geq 6$ can be packed with nets of the cube leaving at most 12 cells uncovered.*

Proof. Figure 6 shows such packings found using BurrTools. □

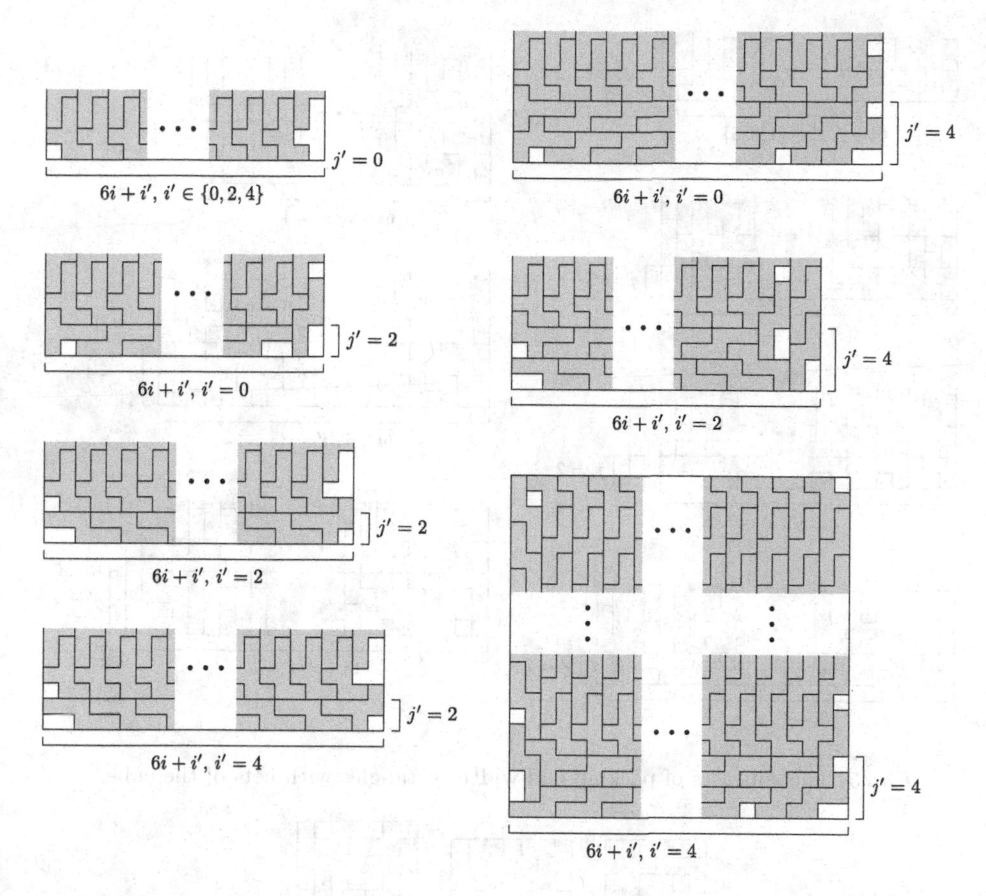

Fig. 4. The subcases of packing even-width rectangles with nets of the cube.

Lemma 3. *Rectangles of dimensions $m \times n$ with $m \geq 6$ and $5 \geq n \geq 2$ can be packed with nets of the cube leaving at most 12 cells uncovered.*

Proof. Figure 7 shows packings for each subcase of $m \geq 6$. In each subcase, at most 5 cells are uncovered at the right end of the rectangle, and at most 7 cells are left uncovered at the left end of the rectangle, for a total of at most 12 (< 14) cells uncovered. □

Lemma 4. *Rectangles of dimensions $m \times n$ with $5 \geq m \geq n \geq 2$ can be packed with nets of the cube leaving at most 10 cells uncovered.*

Proof. The cases $\{2, 3, 4, 5\} \times 2$ and $3 \times, \{2, 3\}$ have area $\leq$10, so the empty packing suffices. The remaining cases are $4 \times \{3, 4\}$ and $5 \times \{3, 4, 5\}$. Refer to Fig. 1. Any single net other than C fits in a 4×3 rectangle and occupies 6 squares,

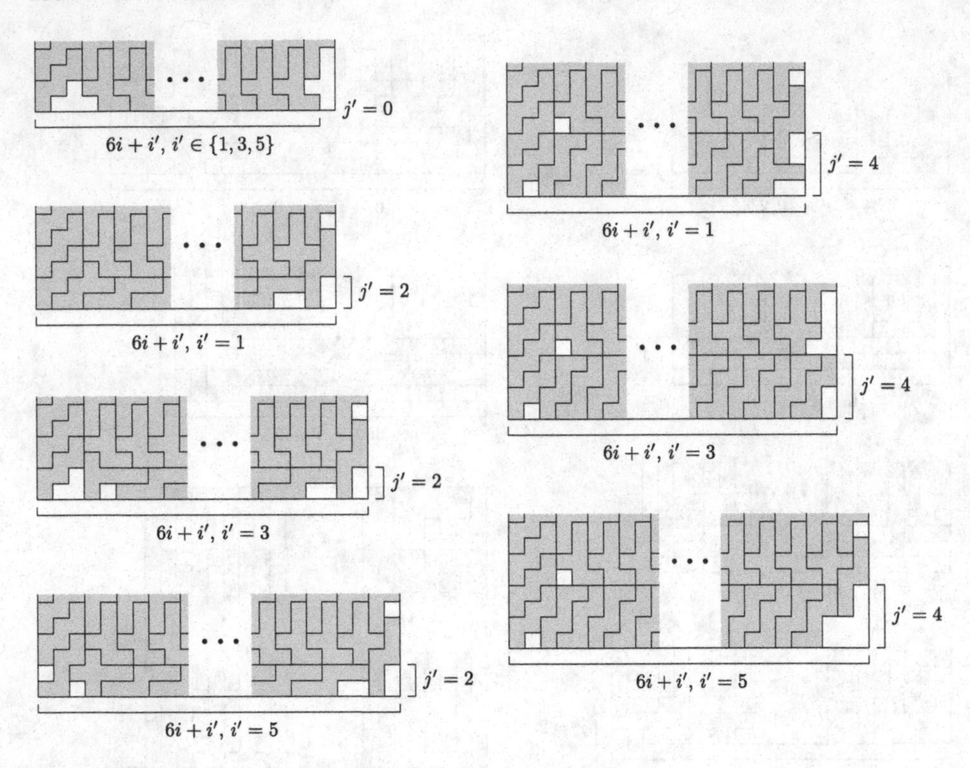

Fig. 5. The subcases of packing odd-width rectangles with nets of the cube.

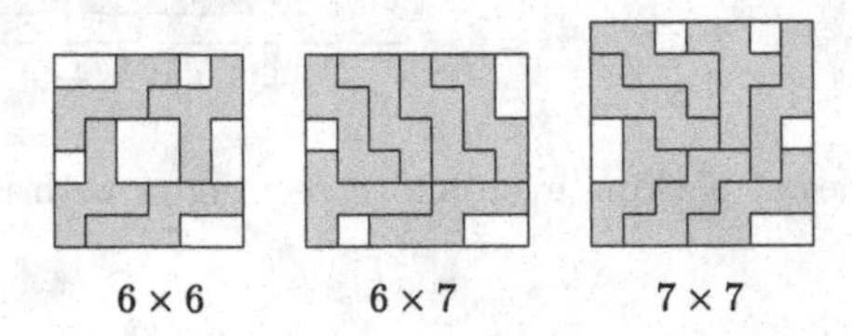

Fig. 6. Packings of 6×6, 6×7, and 7×7 rectangles by cube nets leaving 12, 6, and 7 uncovered cells, respectively.

which is a sufficient packing for $4 \times \{3, 4\}$ (leaving 6 and 10 cells uncovered, respectively) and for 5×3 (leaving 9 cells uncovered). Case 5×4 can be solved by stacking two C nets vertically (leaving 8 cells uncovered). Case 5×5 can be solved by stacking a C above a D (or A) above another C (leaving 7 cells uncovered). □

The previous four lemmas together establish Theorem 2, that 14 is an upper bound on the number of uncovered cells for $m \times n$ with $m \geq n \geq 2$.

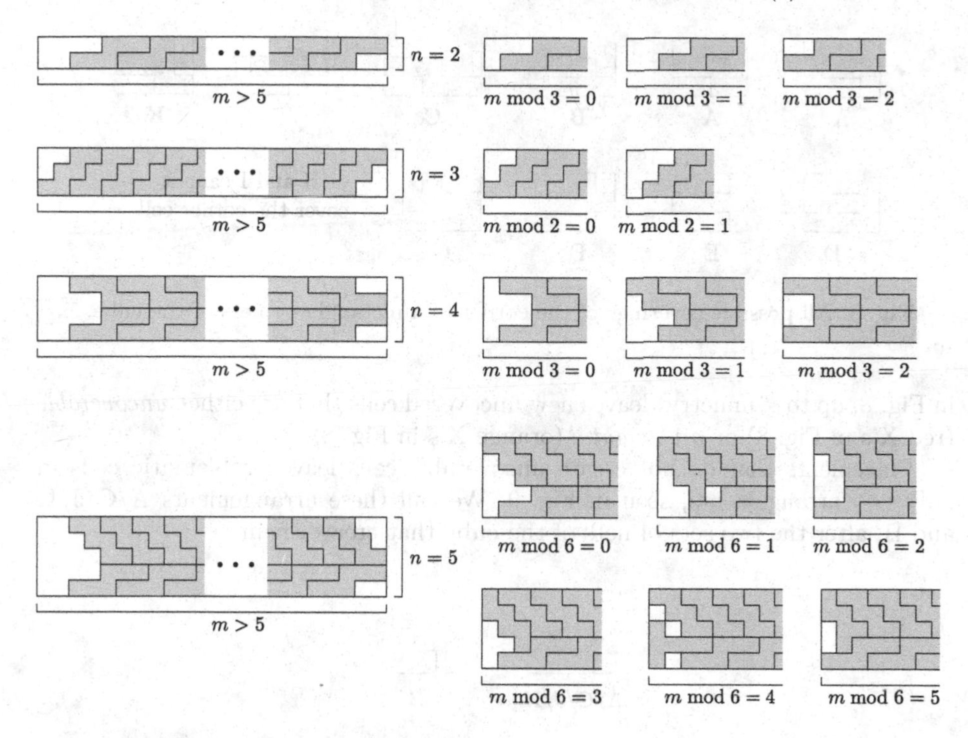

Fig. 7. The subcases of packing rectangles of dimensions $m \geq 6$ and $5 \geq n \geq 2$ with nets of the cube.

3 Lower Bound on Uncovered Cells

In this section, we show the following theorem, which tells us the best case number of uncovered cells when we pack nets of the cube in a rectangle. We cannot do better than leaving 4 uncovered cells for a rectangle of any dimension.

Theorem 3. *Let $m, n \in \mathbb{N}$. If $\{m, n\} \notin \{\{1,1\}, \{1,2\}, \{1,3\}\}$, then any packing of nets of the cube into an $m \times n$ rectangle leaves at least 4 uncovered cells.*

We prove this theorem by dividing it into the following two cases by the sizes of rectangles, that is, (i) 6×2, or larger, and (ii) 5×5, or smaller. Each of these corresponds to one of the subsequent lemmas. Subsequent discussions build on ideas in [12,13].

Lemma 5. *For rectangles of dimensions $m \times n$ with $m \geq 6$ and $n \geq 2$ $(m \geq n)$, any packing of nets of the cube leaves at least 4 uncovered cells.*

Proof. Each of four corners of an empty rectangle is an uncovered cell. We first notice that nets H and I cannot cover a corner cell without protrusion. Observe that any covering of this cell via all possible placements of nets of the cube (seen

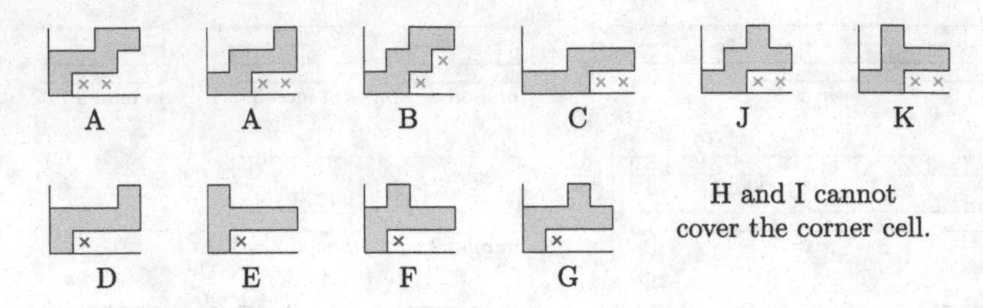

Fig. 8. All possible coverings of the corner of a rectangle by nets of the cube.

in Fig. 8, up to symmetry) leaves new uncovered cells that are either *uncoverable* (red X's in Fig. 8) or *problematic* (orange X's in Fig. 8).

Placements that do not create uncoverable cells leave problematic cells in only two arrangements, seen in Fig. 9. We call these arrangements A/C/J/K and B, after the two sets of nets of the cube that create them.

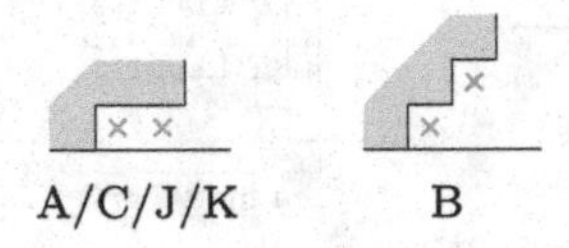

Fig. 9. The two cases of remaining problematic cells after covering the corner of a rectangle by a net of the cube.

For the two problematic cell arrangements A/C/J/K and B in Fig. 9, only four net placements suffice to cover both problematic cells, seen in Fig. 10.

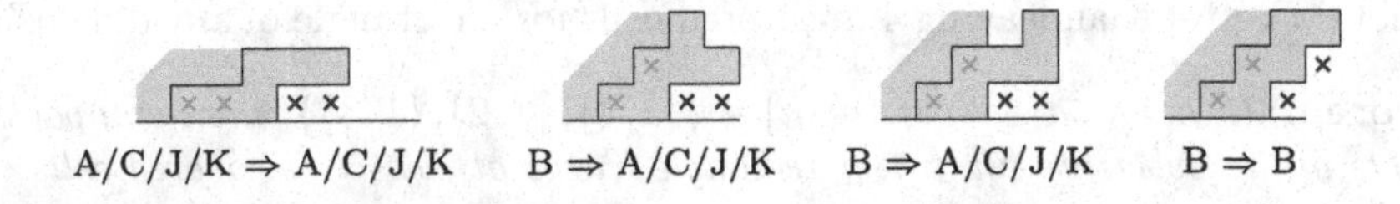

Fig. 10. The four cases of covering both problematic cells (with any number of nets). The blue X's denote new problematic cells created by the new nets.

Note that all four net placements create two new problematic cells either in the arrangement A/C/J/K or B. Thus, covering both problematic cells "propagates" to a new pair of problematic cells in one of the same two arrangements A/C/J/K or B. These problematic cell propagations follow one of the rectangle sides incident to the corner, where the first pair of problematic cells were placed.

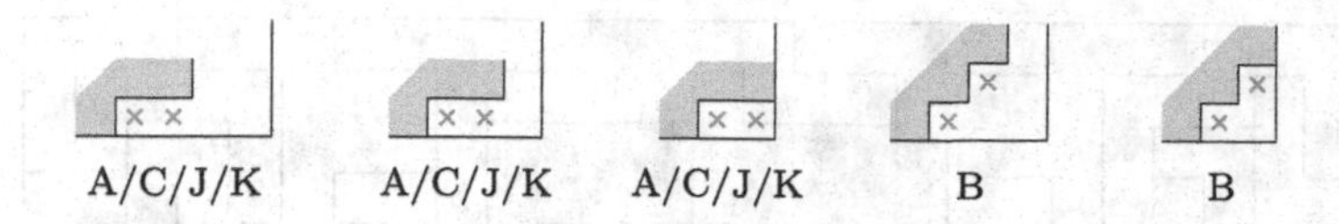

Fig. 11. The cases of a problematic cell arrangement near a corner. In each case, at least two cells are left uncovered.

If such a propagation continues, it eventually reaches another corner of the rectangle (Fig. 11) or a pair of problematic cells propagated from an adjacent corner (Fig. 12). In both cases, at least one of the problematic cells is left uncovered.

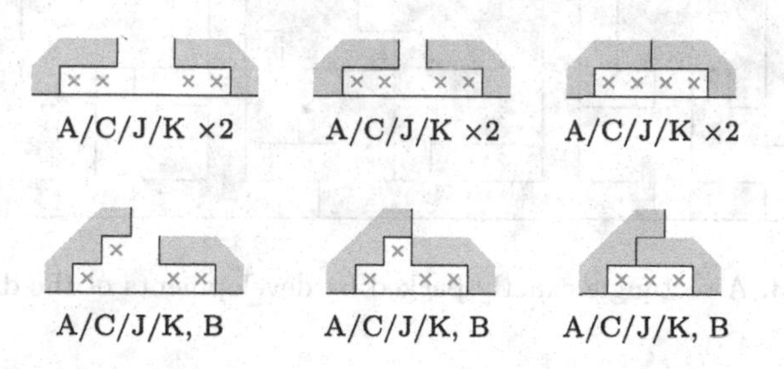

Fig. 12. The cases of a problematic cell arrangement meeting another problematic cell arrangement (each propagated from a different corner). In both cases, at least two of the problematic cells are left uncovered.

In summary, the key observations of the proof are:

- A rectangle has "four" corners.
- Each corner has an uncovered cell or problematic cell arrangement that contains two cells.
- Covering both cells of a problematic cell arrangement creates ("propagates") a new problematic cell arrangement.
- If a problematic cell arrangement is adjacent to a corner, at least one cell is left uncovered.
- If a problematic cell arrangement is adjacent to another problematic cell arrangement (originating at another corner), at least two cells are left uncovered.

Thus at least four uncovered cells are always found in any packing of nets in the rectangle.

Note that the above argument assumed that for each corner, the initial net placed to cover the corner cell created a problematic cell arrangement *attributed to this corner*. If a single net covers multiple corners simultaneously, then this is no longer the case and the proof does not hold. Thus we assume the larger rectangle dimension (m in the lemma statement) is at least 6, implying that

Fig. 13. A rectangle exactly packed by developments of the dicube.

no net can cover non-adjacent corner cells simultaneously. It is easily seen that the only net that is able to cover adjacent corner cells is E (when $n = 3$), in which case, two distinct uncoverable cells, each adjacent to one of the corner cells covered by E, are created. □

We now consider the remaining case.

Lemma 6. *For rectangles of dimensions $m \times n$ with $5 \geq m \geq n \geq 2$, any packing of nets of the cube leaves at least 4 uncovered cells.*

Proof. This is again verified by using BurrTools, seen in Table 1 (left). □

The previous two lemmas together imply Theorem 3, that is, for all rectangles excluding those with area less than 4, any packing of nets of the cube into a rectangle leaves at least 4 uncovered cells, which is a tight bound.

4 Dicube Nets

Now we turn our focus on nets of a *dicube*, that is, (the surface of) a face-to-face gluing of two unit cubes. There is exactly one dicube (up to symmetry), and we can show by enumeration that it has 723 different nets.

Similar to the case of the cube, we ask a primitive question: can any combination of dicube nets exactly pack some rectangle? Surprisingly, in contrast to the cube case, the answer is affirmative.

Theorem 4. *Nets of the dicube can exactly pack a* 26×20 *rectangle.*

Proof. The proof is by demonstration, as we can see in Fig. 13. In total, the packing uses $26 \cdot 20/10 = 52$ nets, which come from 11 distinct nets (up to symmetry). □

This example was found by hand, so we do not know whether it is minimal.

Extending to tricubes (or other n-cubes) is not so simple because there are multiple tricube shapes.

5 Conclusion

We conclude the paper by posing some open questions and conjectures.

As we can observe in Table 1, the number of uncovered cells in all the cases is not greater than 12. We strongly believe that it is upper bounded by 12, and that our upper bound 14 is not tight. Therefore the first open question is the following.

Question 1. Can we prove that the upper bound on the number of uncovered cells is 12?

Furthermore, we believe the following, which partially supports Odawara's conjecture [15].

Conjecture 1. For a $6 \times 6k$ rectangle, the minimum number of uncovered cells is exactly 12 (exactly $6k - 2$ nets are packed).

For nets of the dicube, as we stated in Sect. 4, the following question remains open.

Question 2. What is a rectangle of minimum area that can be exactly packed by nets of the dicube?

Acknowledgments. This research is partially supported by JSPS KAKENHI Grant Numbers JP17K00017, JP21K11757, JP20H05964, JP17H06287, JP18H04091, and JST CREST Grant Number JPMJCR1402.

References

1. Abel, Z., et al.: Unfolding and dissection of multiple cubes, tetrahedra, and doubly covered squares. J. Inf. Process. **25**, 610–615 (2017)
2. Bos, M.: Tiling squares with two different hexominoes. Cubism Fun **70**, 4–7 (2007)
3. Clarke, A.L.: The poly pages: polyominoes: hexominoes, July 2005. http://www.recmath.com/PolyPages/PolyPages/index.htm?Polyominoes.html#hexominoes
4. Dahlke, K.A.: The Y-hexomino has order 92. J. Combin. Theory Ser. A **51**(1), 125–126 (1989)
5. Dahlke, K.A.: Tiling rectangles with polyominoes, November 2017. http://www.eklhad.net/polyomino/index.html

6. Friedman, E.: Polyominoes in rectangles, April 2017. https://www2.stetson.edu/~efriedma/order/
7. Gardner, M.: Polyominoes. In: Hexaflexagons and Other Mathematical Diversions: The First Scientific American Book of Mathematical Puzzles and Games, chap. 13, pp. 124–140. University of Chicago Press (1988). Originally published by Simon & Schuster, 1959, and since republished by MAA/Cambridge, 2008
8. Gardner, M.: Polyominoes and rectification. In: Mathematical Magic Show, chap. 13, pp. 172–187 and 289–292. Mathematical Association of America (1990). Originally published by Knopf, 1977
9. Golomb, S.W.: Checker boards and polyominoes. Am. Math. Mon. **61**(10), 675–682 (1954)
10. Golomb, S.W.: Tiling with polyominoes. J. Combin. Theory **1**(2), 280–296 (1966)
11. Golomb, S.W.: Polyominoes: Puzzles, Patterns, Problems, and Packings. 2nd edn. Princeton University Press (1994)
12. Inoha, T., Inoue, Y., Ozawa, T., Uno, Y.: Packing developments of cubes. In: Abstracts from the 20th Japan Conference on Discrete and Computational Geometry, Graphs, and Games (2017)
13. Inoha, T., Inoue, Y., Ozawa, T., Uno, Y.: Packing developments of cubes: a proof of the theorem. Manuscript (2017)
14. Klarner, D.A.: Packing a rectangle with congruent N-ominoes. J. Combin. Theory **7**(2), 107–115 (1969)
15. Odawara, M.: Development of a cube (in Japanese), September 2008. http://www.torito.jp/puzzles/hako047.shtml
16. Pegg Jr., E..: The eternity puzzle, May 2002. http://www.mathpuzzle.com/eternity.html
17. Putter, G.: Gerard's universal polyomino solver, February 2012. https://gp.home.xs4all.nl/PolyominoSolver/Polyomino.html
18. Reid, M.: Rectifiable polyomino page, October 2005. http://www.cflmath.com/Polyomino/rectifiable.html
19. Röver, A.: Burrtools, October 2018. http://burrtools.sourceforge.net/
20. Selby, A.: Eternity page, May 2006. http://www.archduke.org/eternity/
21. Sicherman, G.: Hexomino oddities, September 2017. https://userpages.monmouth.com/~colonel/6ominodd/index.html

On the Complexity of Jelly-no-Puzzle

Chao Yang(✉)

School of Mathematics and Statistics, Guangdong University of Foreign Studies, Guangzhou 510006, China
http://sokoban.cn/yang

Abstract. Jelly-no-Puzzle is a combinatorial puzzle with gravity and colored blocks. It can be considered as a motion planning problem in an environment in which gravity is taken into account. The computational complexity of Jelly-no-Puzzle is studied in this paper. We show that it is NP-complete for one-color, and is NP-hard in the general case.

Keywords: Combinatorial puzzles · Computational complexity · NP-complete · Motion planning

1 Introduction

Several combinatorial puzzles have the following two features in common: (1) each level of the puzzle takes place in a side-viewed 2-dimensional maze, with the presence of **gravity**; (2) there are several **colored blocks** which can interact with each other.

The game Vexed (also named Cubic) is one of the oldest puzzle of this kind. A level of Vexed is a side-viewed 2-dimensional maze with several colored unit blocks in it, with no adjacent blocks of the same color initially. On the right of Fig. 1, the blocks are distinguished by different patterns instead of colors. The player repeatedly chooses to move a colored block one step to the left or to the right. After the move, all blocks without support fall due to gravity. And after the falling process comes to a stop, any group of connected blocks of the same color disappears. This may again cause a new round of falling and vanishing. The decision problem for Vexed asks whether the player can remove all colored blocks by a sequence of moves. Vexed was shown to be NP-complete [1].

A level of another computer puzzle Clickomania is a rectangular grid full of randomly colored 1×1 blocks. On the left of Fig. 1, a group of connected unit blocks of the some color is shown as a single polyomino, but in fact the unit blocks can fall independently of each other. The player clicks a group of at least two connected blocks of the same color at each step. The selected group of blocks vanishes and the blocks above fall down. The Clickomania problem asks whether the player can remove all the blocks by a number of clicks. It was shown

This work was supported by the Start-up Fund of Guangdong University of Foreign Studies (No. 299-X5219228).

J. Akiyama et al. (Eds.): JCDCGGG 2018, LNCS 13034, pp. 165–174, 2021.
https://doi.org/10.1007/978-3-030-90048-9_13

that the decision problem for Clickomania is in P for one-column grids, and is NP-complete for either 5 colors and 2 columns or 3 colors and 5 columns [2] in 2002. Fifteen years later, it was proved that Clickomania is NP-complete for just 2 colors and 2 columns [3].

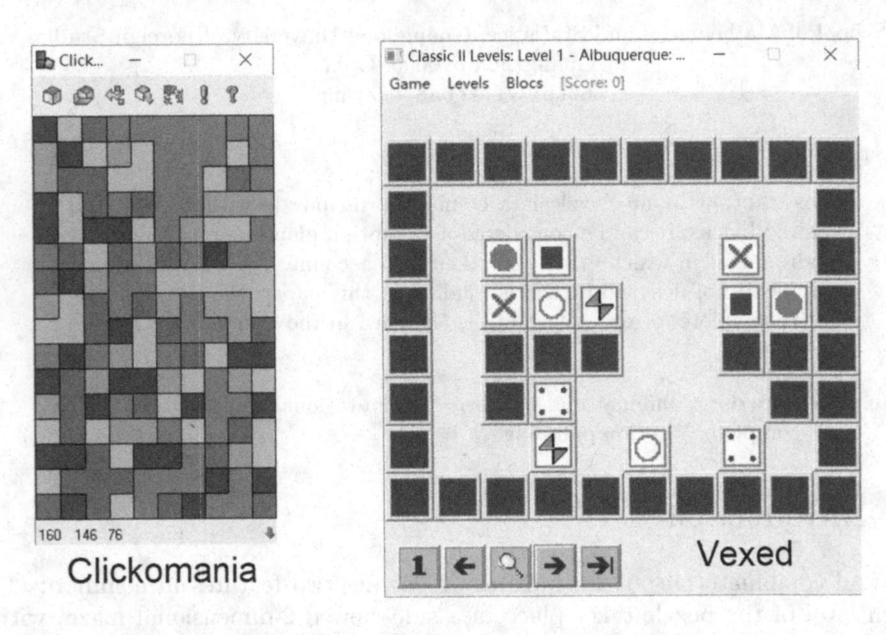

Fig. 1. Screenshot of Clickomania (left) and Vexed (right) (Color figure online)

The goal of both of these puzzles, as well as several variants of match-3 puzzles whose computational complexity was studied in [4], is to eliminate the blocks with the same color or pattern. A box-pushing puzzle with gravity was also shown to be NP-hard [5].

We study the complexity of a puzzle named *Jelly no Puzzle*, which was developed by a Japanese programmer in 2013, and is available for both PC and Android systems[1]. The word *no* in the title of the puzzle is a Japanese word meaning *of*. We shall call this puzzle JELLY for short. The goal of JELLY is kind of opposite to that of Vexed and Clickomania. In Vexed and Clickomania, we try to eliminate all the blocks. But in JELLY, we try to merge all blocks of the same color into one piece (a polyomino). See Fig. 2 for the screenshots of the initial state and a solved state of level 4 of the original JELLY software. The computational complexity of another puzzle with gravity from the same developer of JELLY, the HANANO puzzle[2], has been studied in [6].

Our main result is the following theorem.

---

1 Jelly no Puzzle is available at http://qrostar.skr.jp/en/jelly/.
2 Hanano Puzzle is available at http://qrostar.skr.jp/en/hanano/.

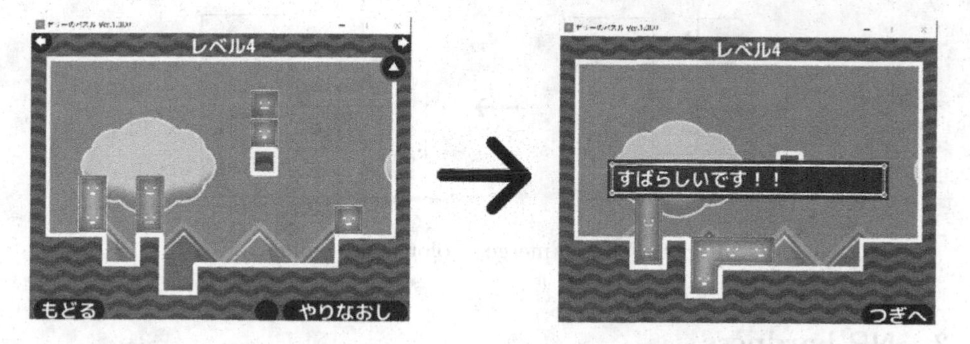

Fig. 2. A level of JELLY with initial state (left), and solved state (right) (Color figure online)

Theorem 1. JELLY *is* NP*-complete for one color, and is* NP*-hard for two or more colors.*

The rest of the paper is organized as follows. The rules of the puzzle are described in Sect. 2. In Sect. 3, we show that JELLY is NP-hard by reduction from the 3-PARTITION problem. In Sect. 4, we show that one-color JELLY is in NP. We conclude by presenting some open problems in Sect. 5.

2 Rules

The rules of JELLY are very simple. Given a maze with fixed platforms and movable blocks of different shapes, sizes and colors, the player can choose to move any block one unit to the left or to the right. After each step, blocks without support will fall due to gravity (Fig. 3).

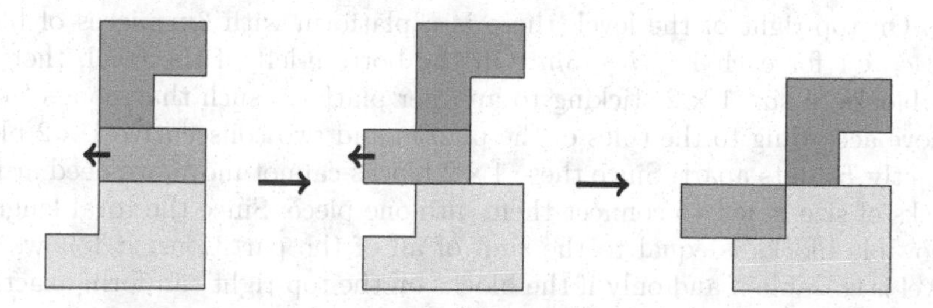

Fig. 3. The red block moves two units to the left, followed by the falling of both blocks (Color figure online)

After falling, two or more blocks of the same color whose edges are adjacent will merge into one block and will remain that way (Fig. 4).

The goal of the puzzle is to merge the blocks of each color into a single block by a sequence of moves. This may resemble some real world applications of motion planning problems in which one has to guide robots of the same type together.

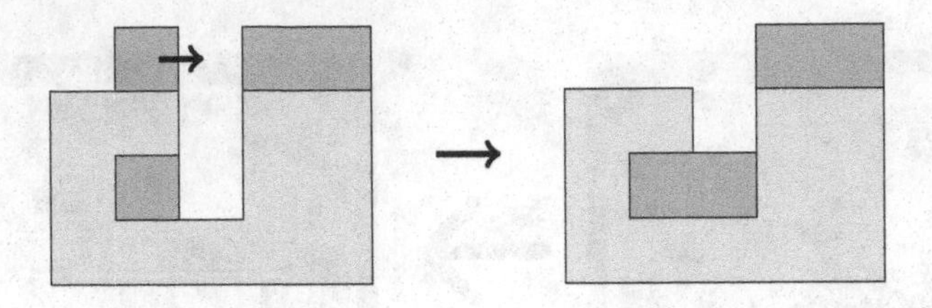

Fig. 4. Fall and merge (Color figure online)

3 NP-hardness

In this section, we prove that JELLY is NP-hard by reduction from the 3-PARTITION problem. The PARTITION problem is one of Karp's 21 NP-complete problems [7,8], and the 3-PARTITION problem has been used to prove the NP-hardness of the classic computer game TETRIS [9,10], which is also a puzzle with gravity.

Definition 1 (3-PARTITION **Problem).**
INSTANCE. *A multiset S of $n = 3m$ positive integers, satisfying $\sum_{c \in S} c = mB$ and $B/4 < c < B/2$ for all $c \in S$.* QUESTION. *Can S be partitioned into m triplets $S_1, S_2, \cdots, S_m$ such that the sums of the integers in each subset are equal?*

Proof of Theorem 1 (NP-hardness). Figure 5 shows a level of JELLY corresponding to the instance of the 3-PARTITION problem in which the multiset is $\{3, 3, 3, 3, 4, 4, 4, 4, 5\}$. In general, given an instance of 3-PARTITION with a multiset of positive integers $\{c_1, c_2, \cdots, c_{3m}\}$ and total sum mB, we will construct a level of one-color JELLY such that the level is solvable if and only if the required partition exists. We assume all blocks in the level to be constructed are in red. On top-right of the level, there is a platform with $3m$ pieces of blocks of size $c_i \times 1$ for each $1 \leq i \leq 3m$. On the bottom-left of the level, there are $m + 1$ blocks of size 1×2 sticking to another platform such that none of them can move according to the rules of the puzzle, and two consecutive 1×2 blocks are exactly B units apart. Since these 1×2 blocks cannot move, we need at least m blocks of size $B \times 1$ to connect them into one piece. Since the total length of the movable blocks is equal to the sum of all of the partitions, it follows that the level is solvable if and only if the blocks on the top-right can form exactly m blocks of size $B \times 1$, which is equivalent to the existence of the required partition of the corresponding instance of the 3-PARTITION problem.

In order to allow the blocks to fall properly onto the correct locations, we also need to place several platforms in the central part of the level. For example, in Fig. 5, two blocks of length 3 and one block of length 5 can form a new block of length 11, so the two blocks of length 3 can fall from the left of platform P_1 and slide to the left, and the block of length 5 can fall from the left of platform P_2. □

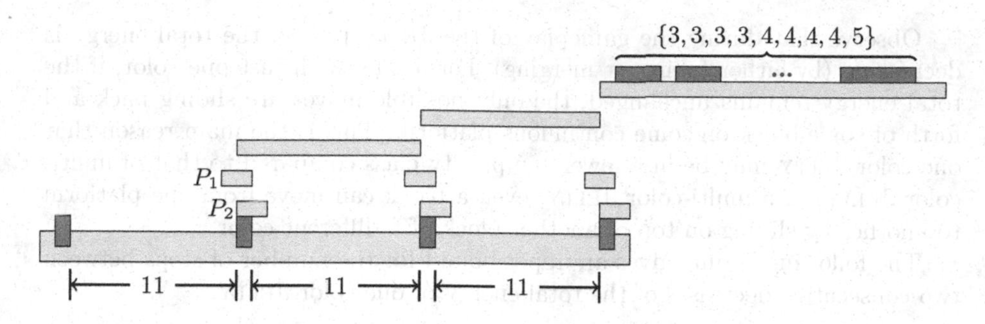

Fig. 5. Reduction from 3-PARTITION

4 One-Color JELLY is in NP

Proof of Theorem 1 *(JELLY∈NP for one-color).* To show that one-color JELLY is in NP, we claim that if a one-color JELLY level is solvable, then it is solvable in a polynomial number of steps relative to the size of the level. Given a configuration C of JELLY level, let $n(C)$ be the number of colored blocks in the configuration C. For each 1×1 square B of some colored block, let $h(B)$ be the height of B in the configuration, and let

$$p(C) = \sum_{B} h(B),$$

where the summation runs over all unit squares of all blocks. Then the total energy of the configuration C is defined by

$$f(C) = n(C) + p(C).$$

The configuration C_0 in Fig. 6 has 4 blocks consisting of 7 unit squares, and it has a total energy

$$f(C_0) = 4 + \sum_{B} h(B) = 4 + (2 \times 3 + 3 \times 2 + 4 \times 2) = 24.$$

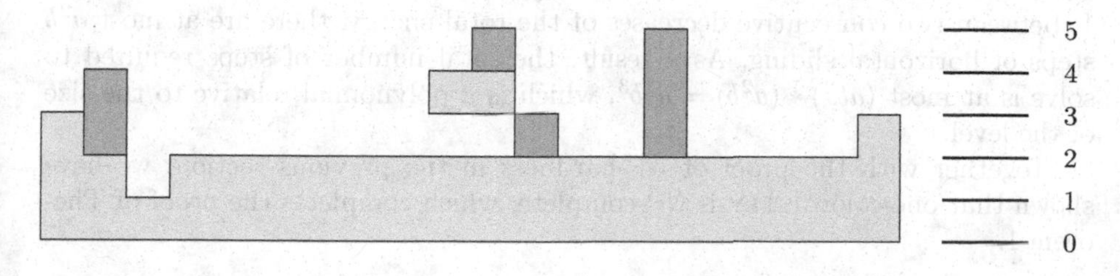

Fig. 6. A configuration with total energy 24 (Color figure online)

Observe that during the gameplay of the JELLY puzzle, the total energy is decreasing (by either falling or merging). For JELLY with just one color, if the total energy remains unchanged, the only possible moves are sliding back and forth of some block on some continuous platform. This is the main reason that one-color JELLY may be in a lower complexity class compared to that of multi-color JELLY. In a multi-color JELLY level, a block can move from one platform to another by sliding on top of another block of a different color.

The following lemma gives an upper bound for the number of steps between two consecutive decreases of the total energy of one-color JELLY.

Lemma 1. *In a one-color* JELLY *level of size $a \times b$, if the configuration C_2 is obtained from the configuration C_1 by sliding of blocks without merging or falling (without changing the total energy), then the number of steps needed for the transition is less than a^2b.*

Proof. We construct a directed graph from the two configurations C_1 and C_2 in the following way. Let the vertex set be the set of blocks whose positions have changed between the two configurations. If a block v_1 cannot slide towards its destination position in configuration C_2 because of the presence of another block v_2, we add a directed edge from v_1 to v_2. Denote this digraph by $D(C_1, C_2)$. See Fig. 7 for an example.

We claim that there is no directed cycle in the digraph $D(C_1, C_2)$; otherwise, all the blocks in the directed cycle cannot move at all without merging, contradicting the hypothesis that we can obtain configuration C_2 from configuration C_1. As a result, there is a block with no out-edge, and this block can move towards its destination by one step. Denote the configuration after this move by C_1'. We can apply the same argument to the configurations C_1' and C_2. By repeatedly sliding blocks in this way, we find a way to change from configuration C_1 to C_2. Note that in each step, the block always moves towards its destination, so the total number of moves of each block is at most a, the width of the level. Since there are less than ab blocks, the total number of steps from configuration C_1 to configuration C_2 is less than a^2b. □

In a one-color JELLY level of width a and height b, the maximum value of the total energy of its initial configuration is less than $ab + ab(b-1) = ab^2$. Since the total energy must be nonnegative, if a one-color JELLY level is solvable, it can be solved by decreasing the total energy at most ab^2 times. And by Lemma 1, between two consecutive decreases of the total energy, there are at most a^2b steps of horizontal sliding. As a result, the total number of steps required to solve is at most $(ab^2) * (a^2b) = a^3b^3$, which is a polynomial relative to the size of the level.

Together with the proof of NP-hardness in the previous section, we have shown that one-color JELLY is NP-complete, which completes the proof of Theorem 1. □

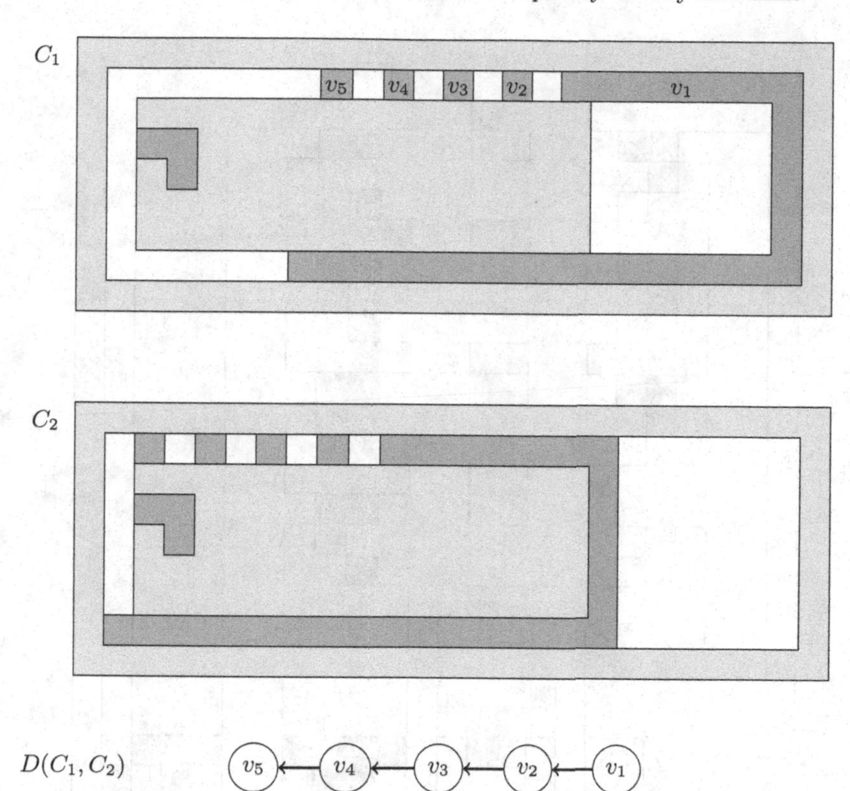

Fig. 7. Configurations C_1, C_2 and digraph $D(C_1, C_2)$ (Color figure online)

5 Conclusion

We have shown that JELLY is in NP for one color. For two or more colors, there are significantly more configurations because blocks of one color can slide on top of blocks of another color. So it is interesting to decide whether JELLY with at least two colors is in NP. One evidence suggesting that the multi-color JELLY may not be in NP is that there exist levels which require an exponential number of steps to solve. By modifying an example in [6], we can construct a family of 3-color JELLY levels whose solutions require an exponential number of steps (Fig. 8). A solved configuration of this level is shown in Fig. 9.

Another related problem is to determine the computational complexity of the no gravity variant of JELLY. In other words, the puzzle takes place in a top-down view 2-dimensional maze, and the puzzle mechanism is otherwise the same. This variant is somewhat similar to the puzzle ATOMIX whose computational complexity has been studied in [11].

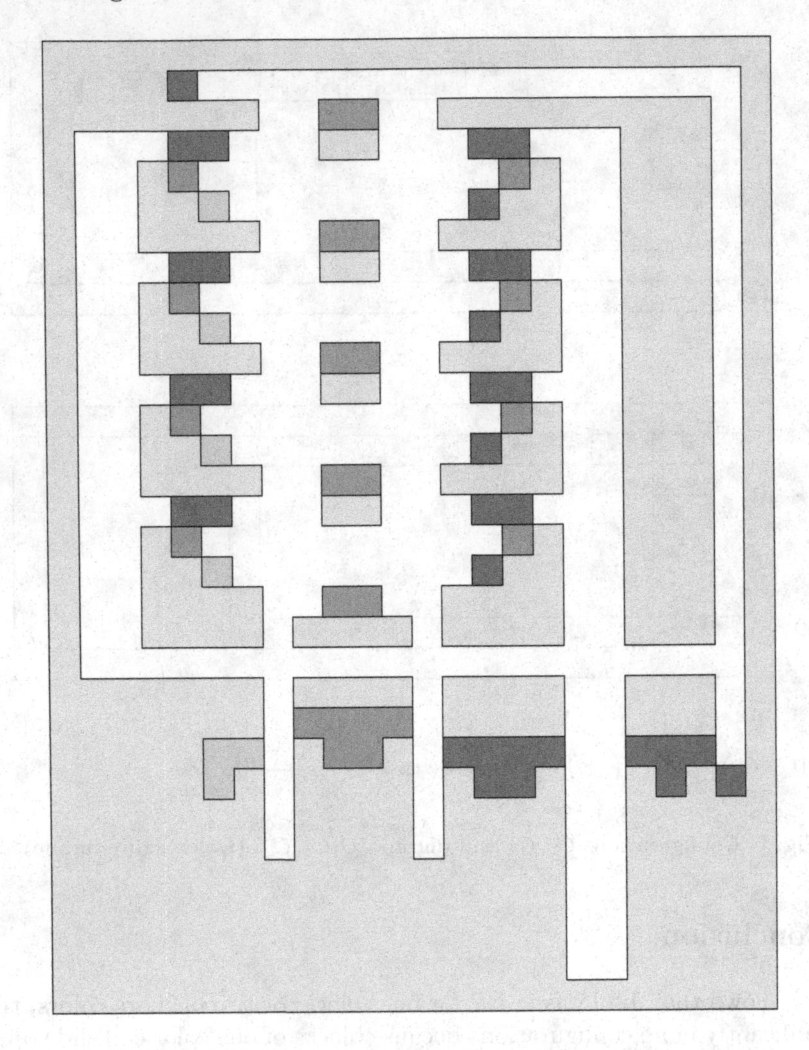

Fig. 8. Example with exponential solution (Color figure online)

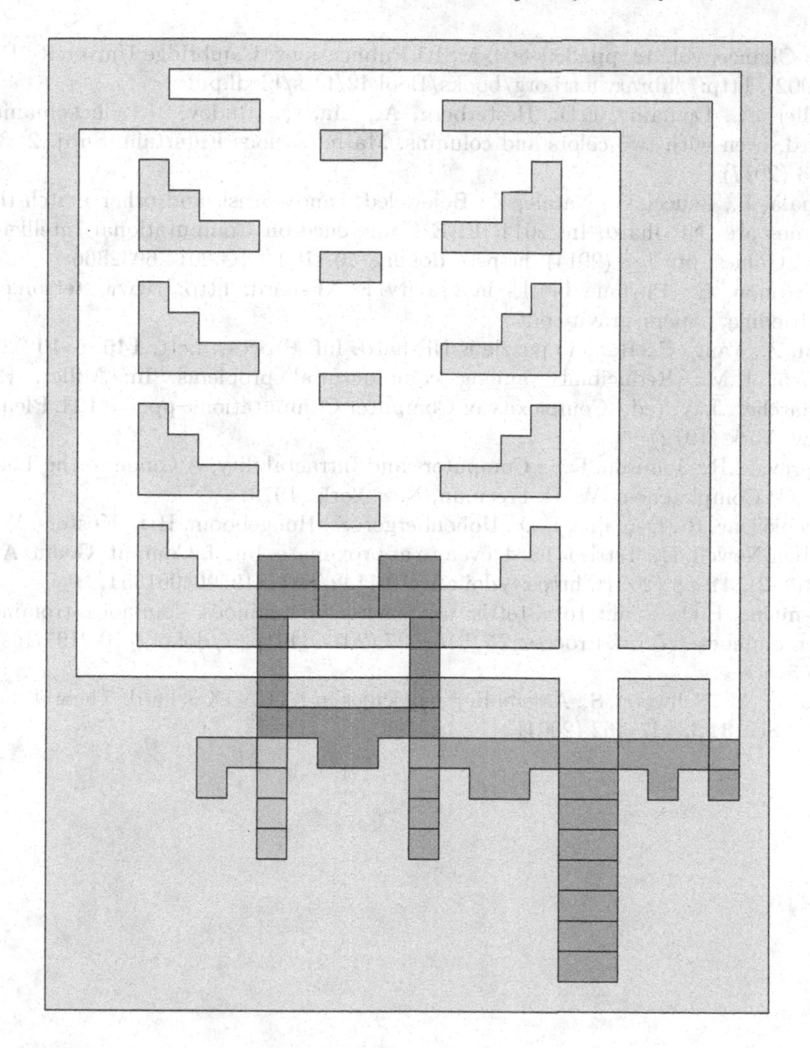

Fig. 9. A solved configuration for the level in Fig. 8 (Color figure online)

Acknowledgements. The author would like to thank the anonymous referees for their very valuable comments which helped to improve this paper, especially for pointing out an error in the original statement of Lemma 1.

References

1. Friedman, E.: The game of cubic is NP-complete. In: Proceedings of Florida MAA Section Meeting (2001). http://sections.maa.org/florida/proceedings/2001/friedman.pdf
2. Biedl, T.C., Demaine, E.D., Demaine, M.L., Fleischer, R., Jacobsen, L., Munro, J.I.: The complexity of clickomania. In: Nowakowski, R.J. (ed.), More Games of

No Chance, vol. 42, pp. 389–404. MSRI Publications, Cambridge University Press (2002). http://library.msri.org/books/Book42/files/biedl.pdf
3. Adler, A., Demaine, E.D., Hesterberg, A., Liu, Q., Rudoy, M.: Clickomania is hard, even with two colors and columns. Math. Various Entertain. Subj. **2**, 325–363 (2017)
4. Gualà, L., Leucci, S., Natale, E.: Bejeweled, candy crush and other match-three games are (NP-)hard. In: 2014 IEEE Conference on Computational Intelligence and Games, pp. 1–8 (2014). https://doi.org/10.1109/CIG.2014.6932866
5. Friedman, E.: Pushing blocks in gravity is NP-hard. http://www.stetson.edu/~efriedma/papers/gravity.pdf
6. Liu, Z., Yang, C.: Hanano puzzle is NP-hard. Inf. Process. Lett. **145**, 6–10 (2019)
7. Karp, R.M.: Reducibility among combinatorial problems. In: Miller, R.E., Thatcher, J.W. (ed.) Complexity of Computer Computations, pp. 85–103. Plenum, New York (1972)
8. Garey, M.R., Johnson, D.S.: Computers and Intractability: A Guide to the Theory of NP-Completeness. W. H. Freeman, New York (1979)
9. Breukelaar, R., Demaine, E.D., Hohenberger, S., Hoogeboom, H.J., Kosters, W.A., Liben-Nowell, D.: Tetris is hard, even to approximate. Int. J. Comput. Geom. Appl. **14**(1–2), 41–68 (2004). https://doi.org/10.1142/S0218195904001354
10. Demaine, E.D., et al.: Total tetris: tetris with monominoes, dominoes, trominoes, pentominoes... J. Inf. Process. **25**, 515–527 (2017). https://doi.org/10.2197/ipsjjip.25.515
11. Holzer, M., Schwoon, S.: Assembling molecules in ATOMIX is hard. Theoret. Comput. Sci. **313**, 447–462 (2004)

Computational Complexity of Two Pencil Puzzles: Kurotto and Juosan

Chuzo Iwamoto[1(✉)] and Tatsuaki Ibusuki[2]

[1] Graduate School of Engineering, Hiroshima University, Hiroshima 739-8527, Japan
chuzo@hiroshima-u.ac.jp
[2] School of Integrated Arts and Sciences, Hiroshima University, Hiroshima 739–8521, Japan
https://home.hiroshima-u.ac.jp/chuzo/

Abstract. Kurotto and Juosan are Nikoli's pencil puzzles. We study the computational complexity of Kurotto and Juosan puzzles. It is shown that deciding whether a given instance of each puzzle has a solution is NP-complete. Each of the two proofs uses a reduction from the PLANAR 3SAT problem.

Keywords: Kurotto · Juosan · Pencil puzzle · NP-complete

1 Introduction

The Kurotto puzzle is played on a rectangular grid of cells (see Fig. 1(a)). Initially, some of the cells contain circles, where each circle contains a number or no number. The purpose of the puzzle is to color cells black according to the following rules [1]: (1) The number in a circle indicates the sum of the number of continuous black cells extending from it, vertically and horizontally (see three cells a, c and d extending from ③ in Fig. 1(c)). (2) Empty circles may have any number of black cells around them. (3) Cells with circles cannot be colored black.

Figure 1(a) is an initial configuration of a Kurotto puzzle. In this figure, there are six circles, five of which contain numbers. From Figs. 1(b)–(f), the reader can understand the basic technique for finding a solution. (b) Consider ② in the red cell. If two grey cells a and b are colored black, then ① in the yellow cell is connected to two continuous black cells. Thus, cells a and c are colored black (see (c)), and cell b must not be colored black; such a cell is indicated by • in Fig. 1. (c) Circled number ③ in the blue cell is connected to black cells c, a and d. (d) Consider ④ in the red cell. If four grey cells containing cell e are colored black, then ① in the yellow cell is connected to four continuous black cells. Thus, cells f and g, h, i, j are colored black (see (e)). One of the multiple solutions is shown in (f).

The Juosan puzzle is played on a rectangular grid of cells (see Fig. 2(a)). Initially, the grid of cells are divided into rectangular *territories*. Each territory contains a number or no number. The purpose of the puzzle is to fill in cells, each with a mark "—" or a mark "|," according to the following rules [2]: (1) The

J. Akiyama et al. (Eds.): JCDCGGG 2018, LNCS 13034, pp. 175–185, 2021.
https://doi.org/10.1007/978-3-030-90048-9_14

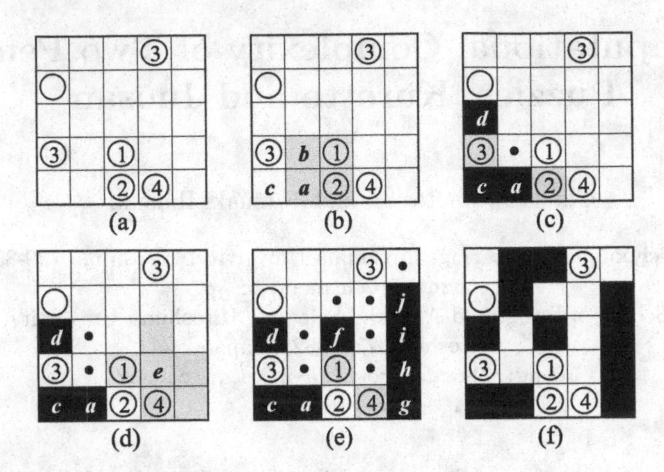

Fig. 1. (a) Initial configuration of a Kurotto puzzle. The progression from initial configuration to a solution is shown in (b)–(f). (Color figure online)

number in a territory shows the number of —-marks if —-marks have majority, or the number of |-marks if |-marks have majority. However, there are also cases where the numbers of —-marks and |-marks are the same. Territories with no numbers may have any number of —-marks and |-marks. (2) —-marks cannot extend to more than two cells vertically. (3) |-marks cannot extend more than two cells horizontally.

Figure 2(a) is an initial configuration of a Juosan puzzle. In this figure, cells are partitioned into ten territories, eight of which contain numbers. In (b), the blue 3×1 territory has number 3, so three |-marks are filled. In (c), if there are three |-marks in the red 2×2 territory, then |-marks extend across three cells horizontally. Thus, the red 2×2 territory will contain three —-marks (see (f)), which implies that the yellow 1×2 territory must contain two |-marks. In (d), since the blue 1×2 territory contains two —-marks, two yellow 2×1 territories contain four |-marks. In (e), since two —-marks are placed in the red 2×2 territory, a —-mark and a |-mark are filled in the blue 1×2 territory. One of the multiple solutions is shown in (f).

In this paper, we study the computational complexity of the decision version of Kurotto and Juosan puzzles. The instance of the *Kurotto puzzle problem* is defined as a rectangular grid of cells, where some of the cells contain circles, and each circle contains a number or no number. The instance of the *Juosan puzzle problem* is a rectangular grid of cells, which is divided into rectangular territories. Each territory contains a number or no number. Each problem is to decide whether there is a solution to the instance. Now we are ready to present our main theorem.

Theorem 1. *The Kurotto and Juosan puzzle problems are NP-complete.*

It is clear that the Kurotto and Juosan puzzle problems belong to NP because, given any proposed solution to an instance of the problem, it can be

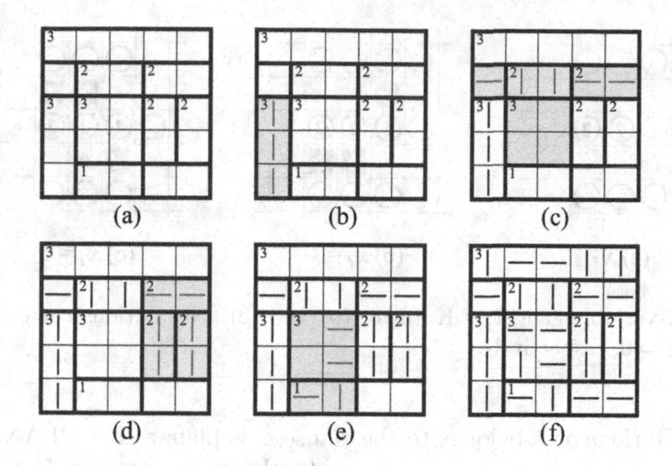

Fig. 2. (a) Initial configuration of a Juosan puzzle. The progression from an initial configuration to a solution is shown in (b)–(f). (Color figure online)

verified in polynomial time if the solution satisfies all the constraints of the problem.

There has been a huge amount of literature on the computational complexities of games and puzzles. In 2009, a survey of games, puzzles, and their complexities was reported by Hearn and Demaine [7]. After the publication of this book, the following Nikoli's pencil puzzles were shown to be NP-complete: Dosun-Fuwari [12], Fillmat [15], Hashiwokakero [4], Herugolf and Makaro [11], Kurodoko [13], Numberlink [3], Pipe Link [16], Shakashaka [5], Shikaku and Ripple Effect [14], Usowan [10], Yajilin and Country Road [8], and Yosenabe [9].

2 NP-completeness of Kurotto

We present a polynomial-time transformation from an arbitrary instance C of PLANAR 3SAT to a Kurotto puzzle such that C is satisfiable if and only if the puzzle has a solution.

2.1 3SAT Problem

The definition of 3SAT is mostly from [6]. Let $U = \{x_1, x_2, \ldots, x_n\}$ be a set of Boolean *variables*. Boolean variables take on values 0 (false) and 1 (true). If x is a variable in U, then x and $\overline{x}$ are *literals* over U. The value of $\overline{x}$ is 1 (true) if and only if x is 0 (false). A *clause* over U is a set of literals over U, such as $\{\overline{x_1}, x_3, x_4\}$. A clause is *satisfied* by a truth assignment if and only if at least one of its members is true under that assignment.

An instance of PLANAR 3SAT is a collection $C = \{c_1, c_2, \ldots, c_m\}$ of clauses over U such that (i) $|c_j| = 3$ for each $c_j \in C$ and (ii) the bipartite graph $G = (V, E)$, where $V = U \cup C$ and E contains exactly those pairs $\{x, c\}$, such

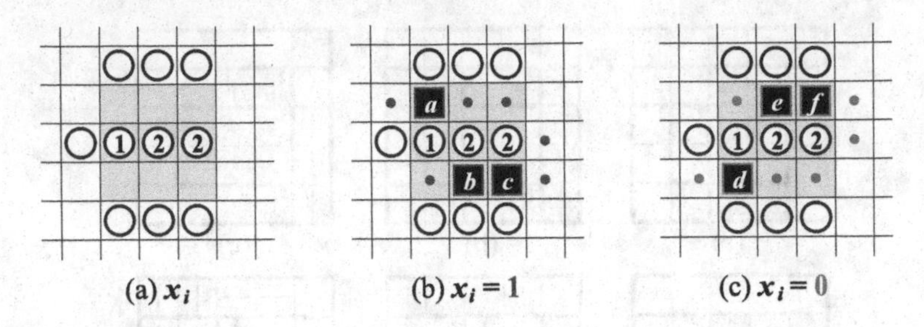

Fig. 3. (a) A variable gadget of Kurotto. (b) and (c) are solutions corresponding to $x_i = 1$ and $x_i = 0$, respectively.

that either literal x or $\overline{x}$ belongs to the clause c, is planar. The PLANAR 3SAT problem asks whether there exists some truth assignment for U that simultaneously satisfies all the clauses in C. The PLANAR 3SAT is known to be NP-complete [6].

For example, $U = \{x_1, x_2, x_3, x_4\}$, $C = \{c_1, c_2, c_3\}$, and $c_1 = \{x_1, x_2, x_3\}$, $c_2 = \{x_1, x_3, \overline{x_4}\}$, $c_3 = \{\overline{x_2}, \overline{x_3}, x_4\}$ provide an instance of PLANAR 3SAT. For this instance, the answer is "yes," since there is a truth assignment $(x_1, x_2, x_3, x_4) = (0, 1, 0, 0)$ satisfying all clauses.

2.2 Transformation from an Instance of PLANAR 3SAT to a Kurotto Puzzle

Variable x_i is transformed into a variable gadget as shown in Fig. 3(a). Figures 3(b) and 3(c) are solutions corresponding to $x_i = 1$ and $x_i = 0$, respectively.

Clause $c_j = \{x_{i_1}, x_{i_2}, x_{i_3}\}$ is transformed into a clause gadget as shown in Fig. 4(a). This gadget is connected to three variable gadgets of x_{i_1}, x_{i_2}, and x_{i_3} (see Figs. 4(b) and 4(c)). Three cells a, b, and c play a key role in this gadget.

If $x_{i_1} = x_{i_2} = x_{i_3} = 0$ (see Fig. 4(b)), then cells a, b, and c must not be colored black. Thus, Fig. 4(b) is an invalid placement of black cells, since the number of continuous black cells extending from ⑲ is 18. Suppose $x_{i_1} = x_{i_2} = 1$ and $x_{i_3} = 0$ (see Fig. 4(c)). If either cell a or b is colored black, then continuous 19 black cells are extended from each ⑲ Now one can see that at least one of variables x_{i_1}, x_{i_2}, and x_{i_3} is 1 if and only if there is a solution to the clause gadget.

Figure 5 is a right branch gadget. The values $x_i = 1$ and $x_i = 0$ are transmitted in the directions of two arrows indicated in Figs. 5(a) and 5(b), respectively. The set of cells in the blue areas of Fig. 5(a) is a right turn gadget. A left branch gadget and a left turn gadget can be constructed similarly.

Figure 6 is a NOT gadget. In Fig. 6(a) (respectively, Fig. 6(b)), if cells a, b, and c (respectively, d, e, and f) in the left grey area are colored black, then cells d, e, and f (respectively, a, b, and c) in the right grey area are colored black.

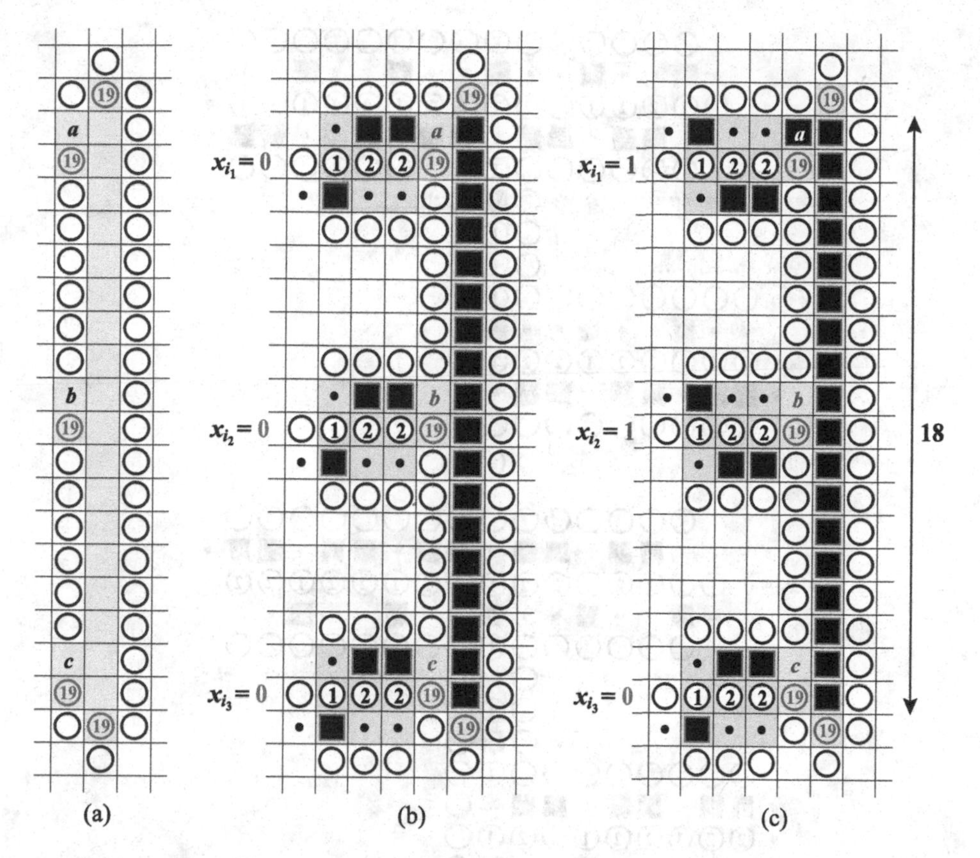

Fig. 4. (a) A clause gadget of Kurotto. Suppose $c_j = \{x_{i_1}, x_{i_2}, x_{i_3}\}$. (b) An invalid placement of black cells, since the number of continuous black cells extending from each ⑲ is 18. (c) If either cell a or b is colored black, then continuous 19 black cells are extended from each ⑲ One can see that at least one of variables x_{i_1}, x_{i_2}, and x_{i_3} is 1 if and only if there is a solution to the clause gadget.

Figure 7 is the top-level description of a Kurotto puzzle transformed from $C = \{c_1, c_2, c_3\}$, where $c_1 = \{x_1, x_2, x_3\}$, $c_2 = \{x_1, x_3, \overline{x_4}\}$, and $c_3 = \{\overline{x_2}, \overline{x_3}, x_4\}$. From this construction, the Kurotto puzzle has a solution if and only if the instance of PLANAR 3SAT is satisfiable.

3 NP-completeness of Juosan

We present a polynomial-time transformation from an arbitrary instance C of PLANAR 3SAT to a Juosan puzzle such that C is satisfiable if and only if the puzzle has a solution.

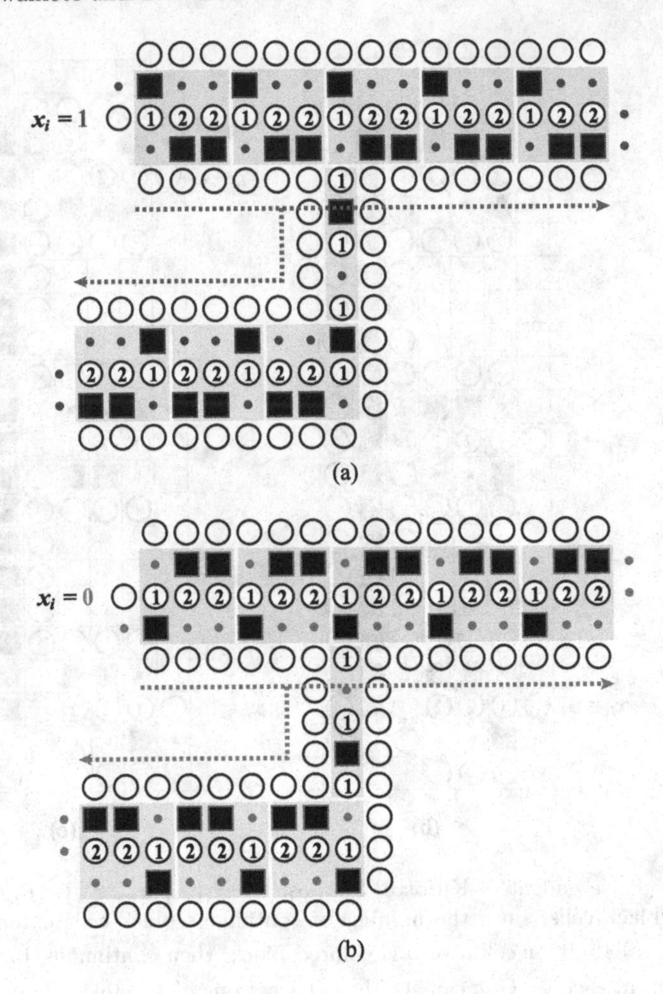

Fig. 5. A Right branch gadget with solutions (a) and (b) when $x_i = 1$ and $x_i = 0$, respectively. The set of connected cells in the blue areas is a right turn gadget.

Variable x_i is transformed into a variable gadget which is composed of 10×10 cells, as shown in Fig. 8(a). Those cells are divided into red and grey cells. The red cells are composed of ten 2×1 territories and ten 1×2 territories, which play a key role in this gadget. In Fig. 8(b) (respectively, Fig. 8(c)), 20 |-marks and 20 —-marks in the blue (respectively, green) area are a solution to the red area of Fig. 8(a). The grey areas of Fig. 8(a) are partitioned into 17 territories. Two of the multiple solutions to those grey areas are given in Figs. 8(b) and 8(c). Figures 8(d)–(f) are simplified illustrations of Fig. 8(a)–(c), respectively.

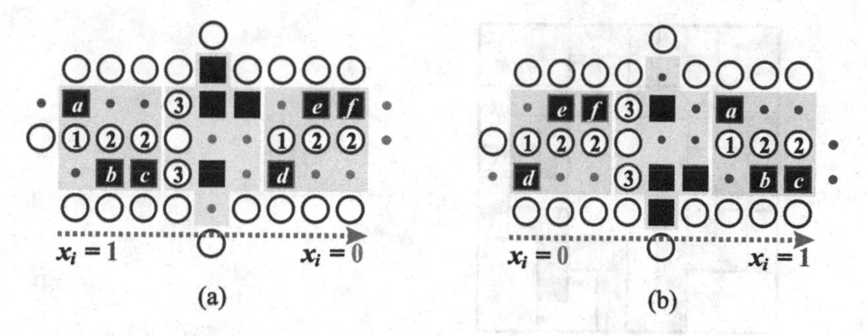

Fig. 6. A NOT gadget. If the left grey area is $x_i = 1$ (respectively, $x_i = 0$), then the right grey area is $x_i = 0$ (respectively, $x_i = 1$).

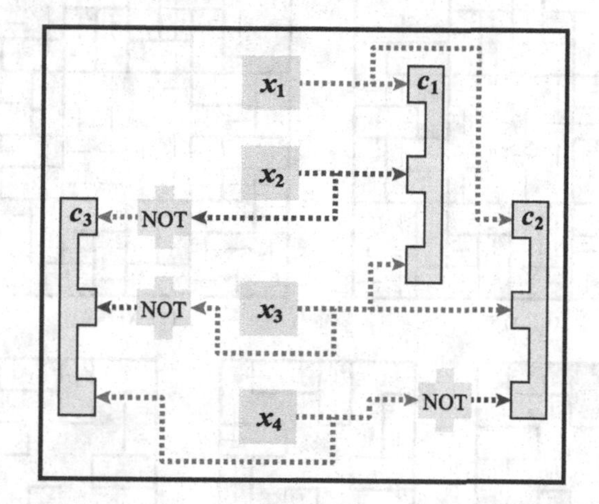

Fig. 7. Top-level description of the puzzle transformed from $C = \{c_1, c_2, c_3\}$, where $c_1 = \{x_1, x_2, x_3\}$, $c_2 = \{x_1, x_3, \overline{x_4}\}$, and $c_3 = \{\overline{x_2}, \overline{x_3}, x_4\}$. From this figure, one can see that $(x_1, x_2, x_3, x_4) = (0, 1, 0, 0)$ satisfies all clauses.

The yellow 41×1 territory of Fig. 9(a) is a clause gadget. This territory contains number 39, and is connected to three variable gadgets of x_{i_1}, x_{i_2}, and x_{i_3} through 12 red territories (see Fig. 9(b)). Suppose $x_{i_1} = x_{i_2} = 1$ and $x_{i_3} = 0$. Cell c must contain a —-mark. If either cell a or b contains a |-mark, then the yellow territory can contain 39 |-marks. Now one can see that at least one of variables x_{i_1}, x_{i_2}, and x_{i_3} is 1 if and only if there is a solution to the clause gadget.

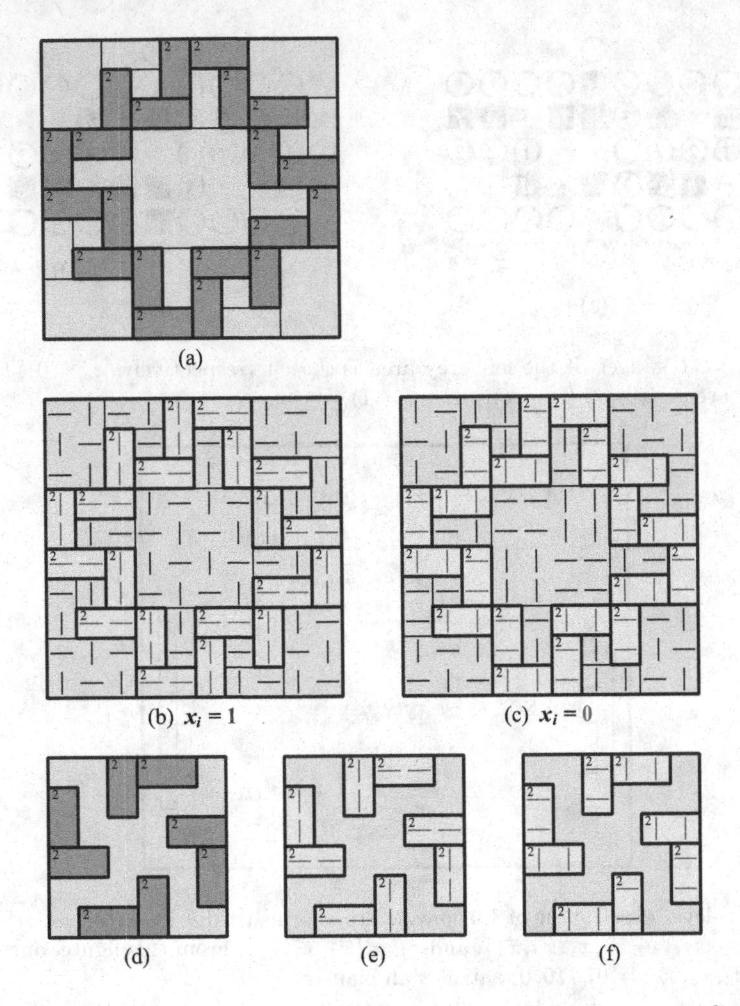

Fig. 8. (a) A variable gadget of Juosan. (b) and (c) are solutions corresponding to $x_i = 1$ and $x_i = 0$, respectively. (d)–(f) are simplified illustration of (a)–(c), respectively. (Color figure online)

Figure 10 is a right branch gadget with a right turn gadget when (a) $x_i = 1$ and (b) $x_i = 0$. Figure 11 is a NOT gadget. In Fig. 11, the left blue area contains the solution for $x_i = 1$ if and only if the right green area contains the solution for $x_i = 0$.

By using variable, clause, branch, turn, and NOT gadgets of Figs. 8, 9, 10 and 11, we can construct a Juosan puzzle of Fig. 7 transformed from $C = \{c_1, c_2, c_3\}$, where $c_1 = \{x_1, x_2, x_3\}$, $c_2 = \{x_1, x_3, \overline{x_4}\}$, and $c_3 = \{\overline{x_2}, \overline{x_3}, x_4\}$. From this construction, the Juosan puzzle has a solution if and only if the instance of PLANAR 3SAT is satisfiable.

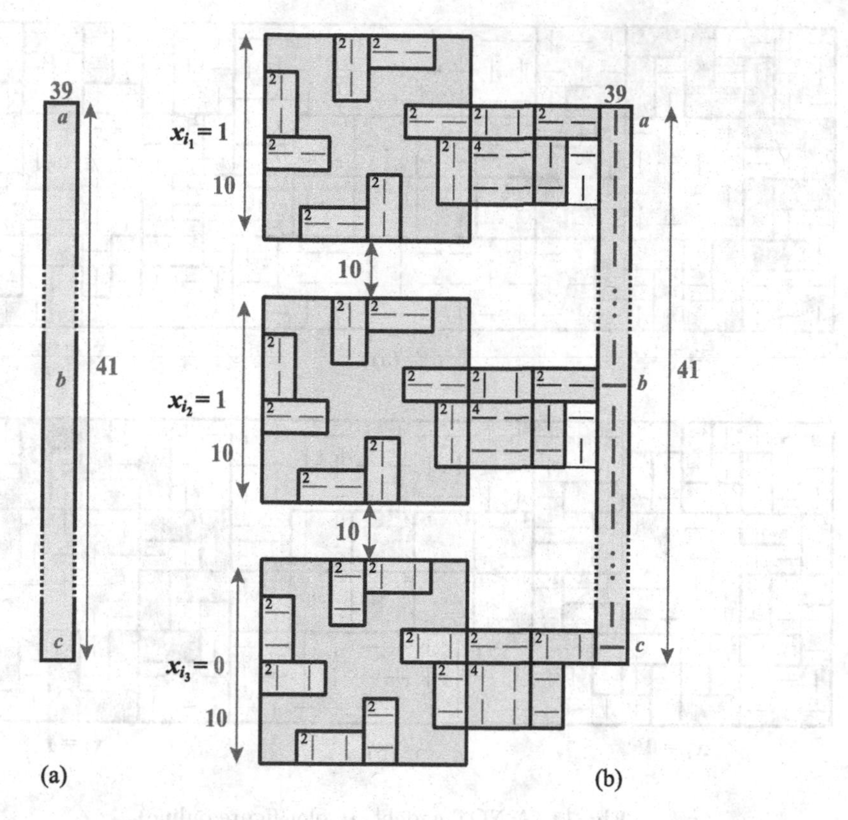

Fig. 9. (a) A clause gadget of Juosan. (b) Suppose $c_j = \{x_{i_1}, x_{i_2}, x_{i_3}\}$, where $x_{i_1} = x_{i_2} = 1$ and $x_{i_3} = 0$. Cell c must contain a —-mark. If either cell a or b contains a |-mark, then the yellow territory can contain 39 |-marks. One can see that at least one of variables x_{i_1}, x_{i_2}, and x_{i_3} is 1 if and only if there is a solution to the clause gadget. (Color figure online)

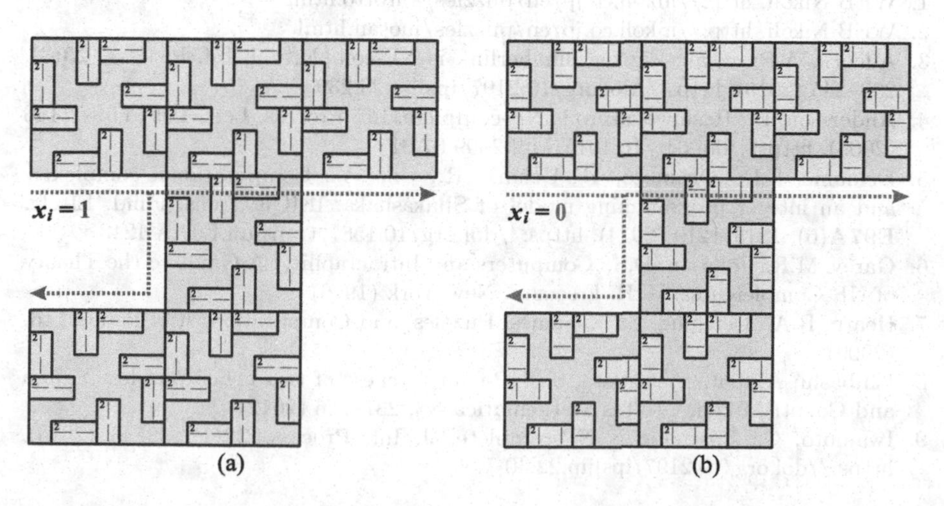

Fig. 10. A right branch gadget with a right turn gadget when (a) $x_i = 1$ and (b) $x_i = 0$.

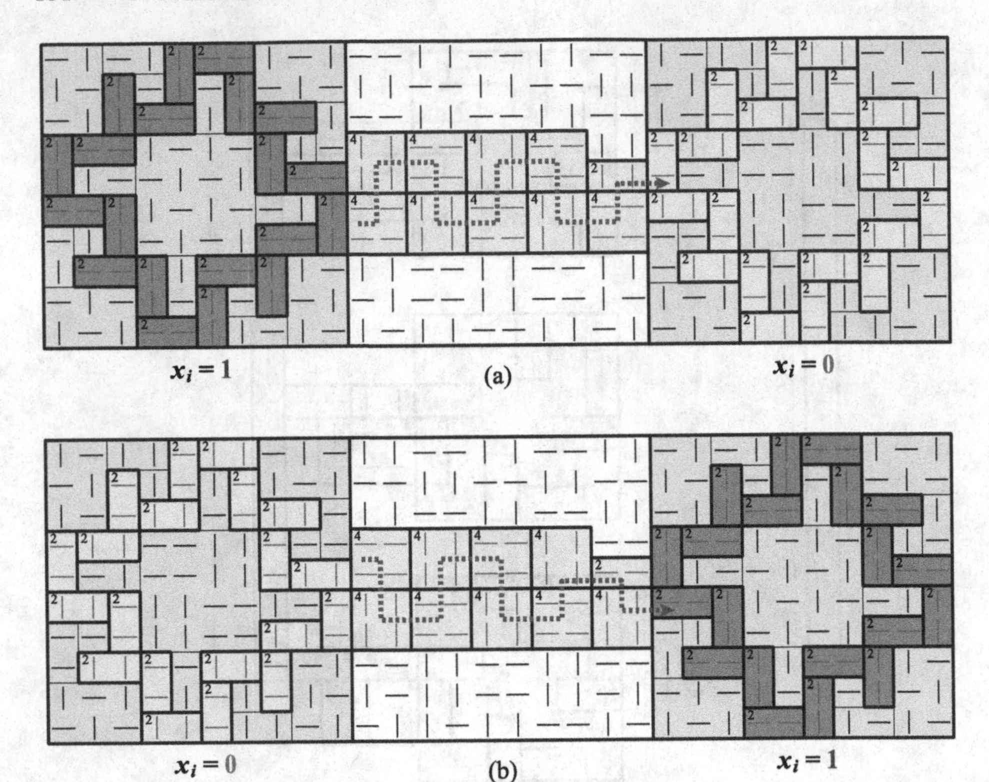

Fig. 11. A NOT gadget. (Color figure online)

References

1. WEB Nikoli. http://nikoli.co.jp/en/puzzles/kurotto.html
2. WEB Nikoli. http://nikoli.co.jp/en/puzzles/juosan.html
3. Adcock, A.B., et al.: Zig-zag numberlink is NP-complete. J. Inf. Process. **23**(3), 239–245 (2015). https://doi.org/10.2197/ipsjjip.23.239
4. Andersson, D.: Hashiwokakero is NP-complete. Inf. Process. Lett. **109**, 1145–1146 (2009). https://doi.org/10.1016/j.ipl.2009.07.017
5. Demaine, E.D., Okamoto, Y., Uehara, R., Uno, Y.: Computational complexity and an integer programming model of Shakashaka. IEICE Trans. Fund. Electr. **E97A**(6), 1213–1219 (2014). https://doi.org/10.1587/transfun.E97.A.1213
6. Garey, M.R., Johnson, D.S.: Computers and Intractability: A Guide to the Theory of NP-Completeness. W.H. Freeman, New York (1979)
7. Hearn, R.A., Demaine, E.D.: Games, Puzzles, and Computation. A K Peters Ltd. (2009)
8. Ishibashi, A., Sato, Y., Iwata, S.: NP-completeness of two pencil puzzles: Yajilin and Country Road. Utilitas Mathematica **88**, 237–246 (2012)
9. Iwamoto, C.: Yosenabe is NP-complete. J. Inf. Process. **22**(1), 40–43 (2014). https://doi.org/10.2197/ipsjjip.22.40

10. Iwamoto, C., Haruishi, M.: Computational complexity of Usowan puzzles. IEICE Trans. Fund. Electr. D101-A(9) (2018). https://doi.org/10.1587/transfun.E101.A.1537
11. Iwamoto, C., Haruishi, M., Ibusuki, T.: Herugolf and Makaro are NP-complete. In: 9th International Conference on Fun with Algorithms. LIPICS, La Maddalena, Italy, 13–15 June 2018, vol. 100, pp. 23:1–23:11 (2018). https://doi.org/10.4230/LIPIcs.FUN.2018.23
12. Iwamoto, C., Ibusuki, T.: Dosun-Fuwari is NP-complete. J. Inf. Process. **26**, 358–361 (2018). https://doi.org/10.2197/ipsjjip.26.358
13. Kölker, J.: Kurodoko is NP-complete. J. Inf. Process. **20**(3), 694–706 (2012). https://doi.org/10.2197/ipsjjip.20.694
14. Takenaga, Y., Aoyagi, S., Iwata, S., Kasai, T.: Shikaku and Ripple effect are NP-complete. Congr. Numer. **216**, 119–127 (2013)
15. Uejima, A., Suzuki, H.: Fillmat is NP-complete and ASP-complete. J. Inf. Process. **23**(3), 310–316 (2015). https://doi.org/10.2197/ipsjjip.23.310
16. Uejima, A., Suzuki, H., Okada, A.: The complexity of generalized pipe link puzzles. J. Inf. Process. **25**, 724–729 (2017). https://doi.org/10.2197/ipsjjip.25.724

Author Index

Printed in the United States
by Baker & Taylor Publisher Services

Printed in the United States
by Baker & Taylor Publisher Services